AUTOMOTIVE ERGONOMICS

DRIVER–VEHICLE
INTERACTION

AUTOMOTIVE ERGONOMICS

DRIVER–VEHICLE INTERACTION

EDITED BY
NIKOLAOS GKIKAS

CRC Press
Taylor & Francis Group
Boca Raton London New York

CRC Press is an imprint of the
Taylor & Francis Group, an **informa** business

CRC Press
Taylor & Francis Group
6000 Broken Sound Parkway NW, Suite 300
Boca Raton, FL 33487-2742

First issued in paperback 2017

© 2013 by Taylor & Francis Group, LLC
CRC Press is an imprint of Taylor & Francis Group, an Informa business

No claim to original U.S. Government works

Version Date: 20120507

ISBN 13: 978-1-138-07315-9

Library of Congress Cataloging-in-Publication Data

Automotive ergonomics : driver-vehicle interaction / [edited by] Nikolaos Gkikas.
    p. cm.
    Includes bibliographical references and index.

    1. Automobiles--Design and construction. 2. Design--Human factors. 3. Automobile driving--Physiological aspects. 4. Human-computer interaction. I. Gkikas, Nikolaos.

TL250.A985 2013
629.2'31--dc23
                2012012861

Visit the Taylor & Francis Web site at
http://www.taylorandfrancis.com

and the CRC Press Web site at
http://www.crcpress.com

Contents

Preface

A NEW PHASE IN THE RELATIONSHIP BETWEEN THE DRIVER, THE VEHICLE AND THE INFRASTRUCTURE

> Car manufacturing cannot turn into a major business, since there are few people who can train as chauffeurs.

> **Karl Benz**

The name of Karl Benz, one of the father figures in the automotive industry, is quoted more than once in this book. This is not only because of his undoubted contribution during the initial phase of automotive development, but also because of the contrast of expectations between key figures such as himself with the established beliefs and practices of today. Common perception of what the automobile is and to whom it is addressed was significantly different back then. From the time when very few could afford it and a handful of those were skilled enough to control such machines, we were led, within a few decades, to the generalisation of the automobile, first in the U.S., then in Europe and post-WWII Japan. Backed by the technological and industrial impetus from two world wars, the automotive industry quickly grew from a niche for the very few, to the ubiquitous supplier of one of the essentials of modern living. The automobile was the ideal vehicle to transfer the technical skills and knowledge developed during the Wars to the relatively peaceful era that followed.

And it did not stop there. The end of the Cold War, the opening of China to the West, the growth of the developing world and the economic boom in the Middle East; all of these events are significant milestones in modern human history, and they all marked new opportunities for further growth in the automotive industry. Openness initially meant opportunities to expand production facilities; quickly, however, production was followed by growth and the emergence of new markets. The automobile has been in the centre of it all along; be it as a commercial product, as a means of transportation, a means of recreation, or an object of art.

Considering all the above and the abundance in technical and non-technical automotive literature, one could argue that the parallel development of vehicle ergonomics has been largely ignored. Significant changes to driver–vehicle interface, such as the establishment of the steering wheel for its biomechanical properties as the de facto control for lateral vehicle control, have largely passed unnoticed. By contrast, there has been a wealth of publications on specific topics such as gearboxes, turbochargers, variable valve-timing or chassis tuning. Against that wealth, there are relatively few, although significant, publications on ergonomics and even fewer books. There was of course the original *Automotive Ergonomics* book from 20 years ago and a few other books on specific areas of research and applications of ergonomics in the automotive domain; still, considering (a) the gravity of the human user as driver, passenger and customer, and (b) the volume of technical information on vehicle attributes with less obvious impact to drivers/passengers/customers, ergonomics is rather scarce.

In addition, recent geopolitical and economic developments such as the emergence of new markets and players, as already mentioned, effectively increased the portfolio of physical, anatomical and cognitive human characteristics that have to be considered during the development of a road vehicle—or any other surface transportation system. Furthermore, the recent technological developments, with the addition of new electronic systems in every vehicle model introduced, set new standards in driver–vehicle interaction, from the moment a customer enters a dealership to examine a prospective vehicle, to the driving experience during the vehicle lifecycle, and the interaction with other road users and facilities in place. It is such developments, socioeconomic on the one hand, technological on the other, that made the present book necessary in the mind of the authors. It is therefore hoped that the pages that follow provide a decent—although imperfect—insight of such phenomena through the eyes of automotive ergonomists to a wider audience.

Nikolaos Gkikas
Autonomics, UK

The Editor

Nikolaos (Nick) Gkikas is a human factors engineer/ergonomist. He holds a PhD in ergonomics (2010) and an MSc in transportation human factors, both from Loughborough University. He is a certified European Ergonomist (Eur. Erg.), a member of the Human Factors and Ergonomics Society (HFES), a member of the Institute of Ergonomics and Human Factors (IEHF), and a founding member and coordinator of the Driving Ergonomics Special Interest Group (SIG), within which the idea of the present book evolved. Dr Gkikas is also a member of the British Standards Institute committees AUE/11, AUE/12 and AUE/14. He has previously worked for the Vehicle Safety Research Group (VSRC) at Loughborough University, which received the Queen's Anniversary Award for Higher Education in 2008 for their contribution to road safety in the United Kingdom. Dr Gkikas has published original research in vehicle HMI, ergonomics and safety. He has also worked as an independent consultant, and as development engineer for Nissan in their European Technical Centre.

The Contributors

Bryan Beeney
Honda of North America
    Manufacturing, Inc.
East Liberty, Ohio

Julie Charland
Dassault Systemes
Montreal, Quebec
Canada

Diane Elizabeth Gyi
Loughborough University
Loughborough Design School
Loughborough, Leicestershire
United Kingdom

Paul Herriotts
Technical Specialist in Ergonomics
Jaguar Land Rover
Gaydon, Warwick
United Kingdom

Simon Hodder
Loughborough University
Loughborough Design School
Loughborough, Leicestershire
United Kingdom

Paul Johnson
Technical Specialist in Vehicle Package
    Engineering
Jaguar Land Rover
Gaydon, Warwick
United Kingdom

Neil J. Mansfield
Loughborough University
Loughborough Design School
Loughborough, Leicestershire
United Kingdom

Andrew Parkes
Transport Research Laboratory
Wokingham, Bershire
United Kingdom

Nick Reed
Senior Human Factors Researcher, TRL
    Senior Academy Fellow
Transport Research Laboratory
Wokingham, Berkshire
United Kingdom

Mark S. Young
Brunel University
School of Engineering and Design
Uxbridge, Middlesex
United Kingdom

1 Automotive Ergonomics 20 Years On ...

Nikolaos Gkikas
Autonomics, UK

CONTENTS

It has been 20 years since the original *Automotive Ergonomics* book, the first comprehensive effort to compile the knowledge of the most important aspects of human–vehicle interaction into a single manuscript. With contributions from some of the most prominent researchers and practitioners of the time, the original automotive ergonomics volume contributed immensely to the propagation of the discipline to both industry and academia across the globe. That volume, edited by Brian Peacock and Waldemar Karwowski (1993), included contributions from a dozen authors, and chapters in occupant packaging, driver vision and visibility, navigation systems and simulation.

Within those 20 years, the automotive industry witnessed an explosion in vehicle electronics and systems development. It is predominantly this rapid development in electronics alongside developments in other areas of vehicle engineering that actually moved the benchmark in applications and standards for automotive ergonomics. Within this framework, and without underestimating the great importance of the original book, it is rather impossible that the original contributions sufficiently cover the ever-developing discipline of automotive ergonomics. A quick look of the contents in the original book alongside the relevant advances in those 20 years will actually shed more light on the argument above.

1.1 THE LINKS TO THE PAST

The chapter on human modelling by Kroemer was one of the pioneering publications of the time; although the basic principles and many of the explicit limitations were laid out in that chapter, the development of technology and the subsequent exponential growth in computer processing power led to the manifestation of readily available human modelling software that could accurately simulate the biomechanics of posture and motion with unprecedented precision. 'Off-the-shelf' packages are nowadays available in the market, and they are accessible to virtually everybody. In parallel, hardware and software capabilities are such that the extent of complexity in the finalised model is only limited by the amount of effort one can dedicate to it, and the accuracy of the original data available to the developer. The inclusion of the physical properties of soft tissue and the growth in the number of joints—and the degrees of freedom in motion—are examples of such evolution. Nevertheless, to a large extent the generic weakness in predicting body motion through the incorporation of behavioural parameters to the models still stands today in the same way it did 20 years ago. In practice, prediction of body motion is very unreliable without the behavioural foundation of motion, that is, the motives for action and the subsequent behavioural patterns that govern motion. Such issues are discussed in Julie Charland's chapter on human modelling and its application within the automotive domain.

Along those lines, vehicle occupant packaging practices benefited from technological developments during the time since the chapter by R.W. Roe was published (1993). Although the general approach of considering individual differences within the sample of the target population remains unchanged, the tools and the details of the methods adopted to achieve the desired occupant accommodation specification has evolved significantly. The evolution of digital human modelling and the aforementioned availability of advanced tools has shifted the practice towards the virtual world in the same manner it made computer aided design (CAD) the norm in industrial design applications. The relevant chapter in the recently published *Ergonomics in the Automotive Design Process* (Bhise, 2011, pp. 29–49) already presented a detailed approach to the process of occupant packaging; Paul Herriotts and Paul Johnson's chapter in the present book provides a fresh view on the matter.

The issue of driver distraction is nowadays one of the pillars of driver ergonomics and safety. With the visual demands being obvious in the common perception of the driving task, a significant proportion of ergonomics research shifted focus from vision and visibility to the study of visual attention and distraction. Night vision, both as driver enhancement through technology and as the study of human vision properties, still retains some momentum (Priez, Brigout, Petit et al., 1998; Källhammer, 2006); however, it is visual distraction that has dominated research irrespective of the scientific, technological or political drives behind it. Thus, the literature is abundant in articles published on that topic. This trend notwithstanding, Haslegrave's chapter in the original book (Haslegrave, 1993) exhibits little effect of aging. Although the contemporary approach nowadays is to include driver visibility as a core occupant packaging practice, in practice very little has changed since Haslegrave's chapter. Visibility remains very much about vehicle-body design, pillar obstructions and windscreen properties. Driver distraction on the other hand, has been driven to a new level. With the introduction of the car/mobile phone, satellite navigation (Sat-Nav), e-mail and Internet facilities, information and entertainment (infotainment) systems in the modern vehicle, more convenience, and more distraction and workload is laid in tandem upon the driver. Nick Reed's chapter in the present book sheds more light on this stone of scandal in modern driver and vehicle safety.

Occupant protection has traditionally been associated with passive safety: the employment of policies and applications which reduce the consequences of an accident. In line with this trend, Lehto and Foley's chapter in the original book (1993, pp. 141–160) was a benchmark summary of the then contemporary knowledge in injury biomechanics and the method by which that knowledge could be applied in vehicle design to mitigate the injury outcome of road vehicle collisions. Nevertheless, the introduction of the antilock braking system (ABS), already in place since the 80s, had paved the way for the family of active safety systems, which today equip every new vehicle in the US, EU and other highly motorized countries. In line with these events, ergonomics research shifted attention towards active safety, accident avoidance and technology that would actually prevent an accident. The next step was the development of advanced driver assistance systems (ADAS), which actively enhance driver control, facilitate the performance of the driving task and ultimately improve the driving experience. Although such technologies are often classified as 'active safety systems' (e.g., collision avoidance systems), they essentially are ADAS systems themselves. Their purpose is to facilitate the performance of the driving task and enhance vehicle control. The continuous development and integration of vehicle systems already available and the addition of newly introduced control systems set the scene for great opportunities, as well as risks. That scene is described and discussed in the chapter by Mark Young in the present book.

One of the biggest challenges in the field of automotive ergonomics was successfully identified quite early by the authors of the original book: aging. In fact, there are two chapters (Imbeau, Wierwille, and Beauchamp, 1993; Smith, Meshkati, and Robertson, 1993) dedicated to the effects of an aging driving population on vehicle design and road safety. The former chapter, (Imbeau, Wierwille, and Beauchamp, 1993) presents a series of psycho-motor experiments that exemplify the sensory decline that comes with age; the latter chapter (Smith, Meshkati, and Robertson 1993) provides the demographic trends of that time (which exponentially extend to our present) and the relevant impact on specific aspects of driver safety. The phenomenon is so widespread throughout the highly motorized world, that its effects are ubiquitous across the range of driver–vehicle interaction—be that vehicle control, the design of displays, In-Vehicle Information Systems (IVIS), seating, occupant packaging or visibility. It is therefore only suitable that every chapter in the present book includes a section or at least considers aging effects in its discussion.

The second big issue nowadays is related to the IVIS systems mentioned above. In the original automotive ergonomics book there was a chapter on navigation, which at the time was making its first steps as a separate entity within the environment of the vehicle. That is one of the areas that has witnessed the biggest developments since that time. The issue is discussed further in Nick Reed's chapter; however, it is worth at this point to note how widespread the use of navigation support nowadays is. A series of both aftermarket and integrated devices is now accessible to virtually every driver. Each and every one of those devices comes with its own features and utilities such as Bluetooth connection, voice control and so forth, and each device is now able to satisfy a variety of drivers with different needs and different abilities.

1.2 NEW OPPORTUNITIES AND THREATS

The original book included some chapters on specific items in vehicle design, such as the design of indirect vision systems (mirrors) and the position of pedals associated with unintended acceleration. The present book does not include such chapters, as they are rather too specific to be addressed within the scope of a single volume. For such specific and, to a certain extent, 'practical' automotive ergonomics issues, the reader is advised to look at Bhise (2011), Porter and Porter (2001), and SAE (1996) ergonomics publications and relevant standards. Instead, the present book includes chapters on topics which either emerged during the last 20 years or are considered to have been somewhat missing from the original book. Those chapters are detailed below.

Chapters on the vehicle as a physical environment: the impact of vibration and thermal effects on vehicle occupants. In traditional vehicle engineering, noise, vibration, harshness (NVH), and heating, ventilation, air-conditioning (HVAC) are two key vehicle attributes specified quite early in the design and development process. Nevertheless, there has so far been limited discussion of the relevant literature from a human-centred perceptive. The following chapters attempt to fill that gap between vehicle attributes and effects on driver and passengers.

Neil Mansfield's chapter discusses vibration; the dynamic environment of a vehicle and how it affects perceptions of quality, causes discomfort, activity interference and occasionally puts health at risk. The chapter provides evidence to support countermeasures and controls for many of the adverse effects of vibration.

Simon Hodder's chapter discusses thermal comfort; ambient heat, radiation, ventilation and humidity are objective parameters that interact between each other and as a group affect thermal comfort. The chapter provides evidence of such interaction as well as the link between objective and subjective assessment of thermal comfort.

Then, there is Diane Gyi's chapter on driving posture and health. The original book from 1993 included a chapter on seating; however, this chapter extends beyond the biomechanical core of seating and approaches driving as a task performed under a variety of conditions and time periods. The latter also includes behavioural parameters and their contribution to posture. In terms of intervention however, the focus is on how an ergonomics intervention program can be applied in the driving environment.

The last completely new topic discussed in this book is the human–machine interface (HMI) in electric vehicles (EV). The epilogue by Brian Peacock in the original book had actually raised the subject as one of the future challenges in automotive ergonomics. That future is now here. Mass-production 'pure' electric and hybrid vehicles are now available worldwide and even hydrogen fuel cell vehicles make their first wheel spin in some parts of the world. Nick Gkikas' chapter in this book describes this very promising world of opportunities (and the risks) from an ergonomics perspective.

The automotive world, both commercially and academically, is among the most competitive and active areas of human occupation. Considering the number of developments in the field and the rate by which they emerge, it is highly likely that this book will also require an update in the not so distant future—and certainly sooner than in 20 years' time. Nevertheless, there has been an attempt to consider anticipated developments in each topic discussed. Only time will tell how successful that attempt has been.

REFERENCES

Bhise, V. D., 2011. *Ergonomics in the Automotive Design Process*. Boca Raton, FL: CRC Press.

Peacock, B. and Karwowski, W. (eds.), 1993. *Automotive Ergonomics*. London: Taylor & Francis.

Haslegrave, C. M., 1993. Visual aspects in vehicle design. In B. Peacock and W. Karwowski (eds.), *Automotive Ergonomics*, pp. 79–98. London: Taylor & Francis.

Imbeau, D., Wierwille, W. W. and Beauchamp, Y., 1993. Age, display design and driving performance. In B. Peacock and W. Karwowski (eds.), *Automotive Ergonomics*, pp. 33–358. London: Taylor & Francis.

Källhammer, J. E., 2006. Night vision: Requirements and possible roadmap for FIR and NIR systems. Available at: http://www.autoliv.com/wps/wcm/connect/fded06004ce4f-41fad53eff594aebdee/SPIE+Paper+6198-14.pdf?MOD=AJPERES.

Lehto, M. R. and Foley, J. P., 1993. Physical aspects of car design: Occupant protection. In B. Peacock and W. Karwowski (eds.), *Automotive Ergonomics*, pp. 141–160. London: Taylor & Francis.

Porter, J. M. and Porter, C. M., 2001. Occupant accommodation: An ergonomics approach. In J. Happian-Smith (ed.), *An Introduction to Modern Vehicle Design*, pp. 233–275. Oxford: Butterworth-Heinemann.

Priez, A., Brigout, C., Petit, C. and Boulommier, L., 1998. Visual performance during night driving. *Proceedings of the 16th Enhanced Safety in Vehicles (ESV) Conference*. Paper Number 98-S2-P-23. Windsor, Ontario: National Highway Traffic Safety Administration.

Roe, R. W., 1993. Occupant packaging. In B. Peacock and W. Karwowski (eds.), *Automotive Ergonomics*, pp. 11–42. London: Taylor & Francis.

SAE, 1996. SP-1155, *Automotive Design Advancements in Human Factors: Improving Driver's Comfort and Performance*. Warrendale, PA: Society of Automotive Engineers.

Smith, D. B. D., Meshkati, N. and Robertson, M. M., 1993. The older driver and passenger. In B. Peacock and W. Karwowski (eds.), *Automotive Ergonomics*, pp. 453–472. London: Taylor & Francis.

2 Digital Human Modelling (DHM) in the Automotive Industry

Bryan Beeney
Honda of North America, US

Julie Charland
Dassault Systemes, Canada

CONTENTS

2.1 INTRODUCTION TO DHM: HISTORICAL OVERVIEW

The concept of a mathematical man-model appears feasible and future efforts should continue to refine and improve the model as well as the validation criteria and methods.

7

TABLE 2.1
Non-Exhaustive Digital Human Modelling (DHM) Tools List

DHM Name	Owner	Characteristics
Anybody[a]	Anybody Technology	Musculoskeletal modelling software
Human[b]	Dassault Systemes	Virtual ergonomics in PLM for multiple industries in design, digital manufacturing and digital maintenance
Jack[b]	Siemens	Human modelling in PLM for multiple industries in design, digital manufacturing and digital maintenance
Madymo[a]	TASS	Generic multi-body finite element software for impact simulation
Ramsis[a]	Human Solutions	Human modelling in occupant packaging mainly used in automotive
Santos[a]	Santos Human	Human modelling mainly funded and used by the US Army

Note: An extensive review of DHM history as well as DHM tool description can be found in Chaffin, D. B., 2001, *Digital Human Modelling for Vehicle and Workplace Design*, Warrendale, PA: Society of Automotive Engineers. A more recent review of DHM tools is also presented in Bubb, H. and Fritzsche, F., 2009, A Scientific Perspective of Digital Human Models: Past, Present and Future, In *Handbook of Digital Human Modelling, Research for Applied Ergonomics and Human Factors Engineering*, V. Duffy (ed.), Boca Raton, FL: Taylor & Francis.

[a] Standalone product.
[b] Product integrated in a PLM Solution.

This sentence was used in a report released in the late 1960s by Boeing's military department (Ryan and Springer, 1969). The program objective was to develop the first known digital human model to simulate motions of any sized human operator at any particular workstation. Even if DHM seems to be quite recent for some people, this reference confirms that it has been around for a while.

It was following these pioneers' work that research on DHM tools started (Table 2.1). Hence, other organizations followed Boeing's footsteps and DHM tools began to appear. This is an example of the birth of a DHM tool: Genicom Consultants Inc., a consulting ergonomics firm in Canada, was founded in 1984 by a group of professionals, including a professor of École Polytechnique de Montréal. In the early 90s, an internal decision was made to develop a DHM product called 'Safework Pro' to speed up consulting projects as well as distribute licenses for professional use in design and virtual manufacturing.

In December 1999, Genicom Consultants Company (renamed Safework Inc.) became a wholly owned subsidiary of Dassault Systemes who provides product life-cycle management (PLM) solutions. As a result of this agreement, Safework became Dassault Systemes' human modelling and virtual ergonomics competency center to provide computer-aided design–computer-aided manufacturing (CAD–CAM) environment-integrated virtual ergonomics solutions.

Over the years, DHM therefore went from a research context usage to various industrial applications. It also went from using templates (such as human scale) to 2D manikins (such as the SAE standard J826 H-point manikin), and then to a complete 3D human model.

DHM is used to ensure that ergonomic considerations will be taken into account as soon as possible during the design process in a 3D environment. It allows for design changes when those are easy to make and inexpensive. Hence, in a 3D environment, changing a car interior setup or a manufacturing task is much more feasible than in a real context, once everything is in use or into production. One can probably easily imagine the difference in cost when making changes to a workstation on a car assembly line while in 3D, in comparison to making those changes by having to stop the production line.

2.2 DHM APPLICATIONS IN THE AUTOMOTIVE INDUSTRY

DHM is used most widely in the automotive industry. The Society of Automotive Engineers (SAE) has participated in promoting both research and conferences through the SAE G-13 committee as well as in organizing a yearly event known as the DHM conferences until 2008. Since then, DHM conferences are held either independently or embedded within more generic ergonomics events such as the International Ergonomics Association (IEA). Automotive was the first industry to design a car entirely in 3D before the first real part even existed. This represents an example of DHM adoption within a product lifecycle management (PLM) environment. DHM is also used in the aerospace industry for cockpit design, digital manufacturing and maintainability. A simple example of digital maintainability in aerospace is the study of analyzing if an airplane engine can be safely maintained by the targeted people and in a timely manner. This last consideration is quite important knowing the high cost of leaving an airplane on the ground for repair when not planned. Ergonomic assessment of digital maintenance leads to product changes early in the design process. Other industries such as industrial equipment, shipbuilding and energy are now tackling DHM as well. A good example related to the energy industry is that of refurbishing: when closing a nuclear plant for 2 weeks, the cost of non-operation is quite high. Planners need to ensure that all operations can be achieved in a safe and timely manner. The main DHM applications in automotives are vehicle occupant packaging and digital manufacturing.

2.2.1 VEHICLE OCCUPANT PACKAGING

Cockpit design and layout are driven by usability, ergonomics, comfort, visual presence, technology, and so forth. These studies can be performed early in the design process using DHM, without having to build physical mockups and to test with real people representing the targeted population.

Designers use these studies to understand how people will use or move around in the product. Essentially, analysis would help in understanding how the buttons and commands work in conjunction with the driver's action as well as how the driver 'fits' in the interior and seating environment. Analysis can entail postures, reach, space, forces, comfort and visual surroundings.

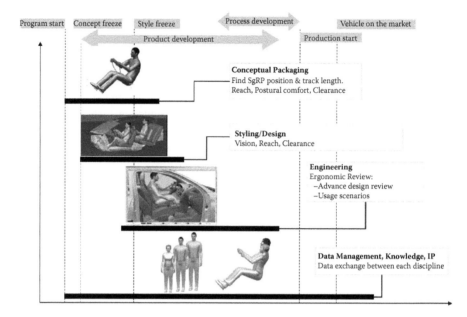

FIGURE 2.1 While each automotive company most probably has specific requirements to cover for their targeted users, this is a generic occupant packaging process.

Using DHM provides the strength of having all domain actors who collaborate during the design process: seat provider, dashboard designers, members of the structures team, and any other stakeholders. All these people can work in parallel taking into consideration changes introduced by others as the design evolves (Figure 2.1).

While not completely possible today, designing only in 3D is feasible. One important issue is related to perception: there is a huge variability between all the potential users. As we look at some areas that can cause newer DHM users concern, the purely visual aspect of skin compression becomes noticeable (pressing a button or a switch, holding an item or being in a seated posture). In addition, clothing, drape, compression and restriction are items that begin to appear as the user gains a more knowledgeable understanding of the modelling tool. Adding this new discovery to an understanding of how the human factor actually functions drives more questions. But the positive to these concerns is being researched: skin wrap, soft tissue complexity and clothing movement and restriction are on the horizon. In context, this still appears to be mainly visual purpose-driven research, but to begin real analysis, it is needed to understand how each of these issues impacts reaches, forces and postures.

More research is also still needed to confirm cognitive load, stimulus response and psychological aspects. These all lead to the human factor of response timing. A canned time and adjustments can be loaded into the DHM; but a deeper understanding is required to commit to virtual simulation verses actual results. Safety is the most important aspect in this design process, so making the leap totally to virtual is yet to come.

2.2.2 Digital Manufacturing

DHM solutions are used mainly by ergonomists, industrial engineers and manufacturing engineers in specific digital manufacturing processes such as: component or tool reach capability, workstation layout, assembly sequence validation and line balancing.

The use of DHM creates a virtual environment that allows the users to present analysis results to a wide range of experienced members such as parties from manufacturing employees to top management coming from separate disciplines or responsibilities within an organization. This context makes it possible to collaborate on a single important aspect of a design up to multiple areas of needs. As businesses move forward and are challenged to produce better products more rapidly, this activity becomes essential for decision-making processes.

Another benefit of using DHM in virtual manufacturing is the ability to study new products or environments that still only exist at design stages. Having that opportunity to interact with the 3D data gives an opening for feedback that could lead to requesting some changes or modifying the design to lower work-related injuries and ergonomic stresses (Figure 2.2).

By simulating assembly tasks with the use of digital humans, DHM users try to gain an understanding of what a part, a product or a workstation will be like by comparing it to something similar in today's world. For example, how will the part or product be installed? How will the parts be delivered to the workstation? How will the employee move and work in that station? Of course, the success of the human factors analysis is directly linked to the acceptance and real results by the employee himself. These are accomplished through static posture and simulations based on the details needed by the user.

Probably the most critical piece of the virtual analysis is validation. Going on an actual follow-up to the part, product or workspace, evaluating the situation and comparing 'actual results to the virtual plan' is vital to improving the methodology in 'how to' analyses for future studies. Without this step, progress can be slower than expected. Plus, the user loses the benefits of understanding in what areas the DHM

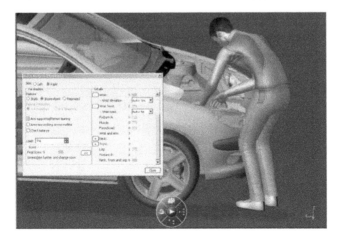

FIGURE 2.2   Virtual ergonomics in digital manufacturing.

lacks the expected responses. This increases the risk of not knowing what research is needed and losing opportunities for appropriate funding and future investments. Very few DHM validation studies have been conducted, so the one done by Oudenhuijzen, Zehner, and Hudson (2009) is quite valuable.

2.3   GENERIC DHM WORKFLOW

2.3.1   Load 3D Environment Data

One of the added values of using DHM is to be able to predict postures as well as to evaluate ergonomic stresses in 3D. In order to do so, the parts that the manikin representing a worker, an operator or a user will be interacting with must be part of the 3D environment (Figure 2.3).

2.3.2   Position/Posture Manikins

Once the 3D environment in which the manikin(s) will be interacting with is complete or at a level of detail that allows for proper analysis, one or more manikins need to be positioned at the right place. When assessing an ergonomic review for the first time, manikin(s) can manually be placed at the right location or automatically be placed using different tools such as pre-set manikin positions saved in a catalog or snap tool.

2.3.2.1   Manual Posturing

The manikin then needs to be postured either to get static posture or with the objective of creating a simulation. This can be done manually, using the mouse and manipulating the manikin and its segments with different types of positioning tools such as forward kinematic, inverse kinematics, segment degree angle edition, and so forth. In some cases, for tool manipulation, for example, where space reservation for a hand is needed, a digital forearm could be used.

2.3.2.2   Automatic Posturing

Manual posturing is quite acceptable for basic posture and in a context where precision is not needed. However, when complex postures are involved as well as back

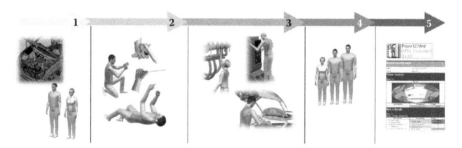

FIGURE 2.3   Here is a workflow that has been simplified and generalised to support several types of DHM use so that the reader can easily map it to his domain.

implication such as bending, rotation and asymmetric position, tied with precise analysis such as force, fatigue or time, then an automatic posturing using posture prediction methods is highly suggested. This allows the manikin to adopt a realistic posture based on mathematical equations derived from specific studies. Several labs have worked on such techniques. As an example, the Human Motion Simulation (HUMOSIM) Laboratory develops data-grounded models to predict and evaluate realistic human movements (Reed, Faraway, and Chaffin 2006). Such posture prediction tools mixed with collision avoidance technology such as the one developed by the Kineo Computer Aided Motion software company provides an efficient way of building simulations.

2.3.2.3   Positioning with Motion Capture Systems

Another way of positioning the manikin is using virtual reality (VR) devices such as motion capture systems and cybergloves. Such devices allow for a real user to move and replicate the motions on the 3D manikin exactly as they are done on the real human. This can be done in a fully immersive environment, meaning that the real person is wearing a helmet to see what is in the 3D environment. A mix environment can also be made using some parts that are real. An example would be to simulate climbing on a platform or crawling into a tight space. Another technology that is quite useful in an assembly/disassembly context is the use of a haptic device. When the real person manipulating a part (a hand on the haptic device arm) makes a collision in 3D, he can actually feel the collision and even feel the stop on collision. This allows for finding a feasible path. Today's high-end VR systems are used mainly by original equipment makers (OEMs) and research centers. Price and complexity are the main consideration in spreading to small- and medium-size businesses (SMBs). In such cases, Microsoft Kinect might be a good alternative but more evaluations are needed.

2.3.3   PERFORM ANALYSES

Analysis can be done on static posture or on simulation. Best practices involve checklists as a baseline to validate overall time, accessibility, reach ability, vision, space, safety, fatigue and strength. Thus, considering many items found in design are related to some checklist but many new designs are introduced or modified where a problem had not existed before. Understanding how a product is made in its environment, or how it will be used, are keys to finding changes that may benefit the most in the analysis phase.

2.3.4   ANTHROPOMETRY

Once the 3D environment is available and the manikins have been positioned at the right place with proper posture and simulation when needed, the next step for a DHM user is to ensure that the postures and simulations can be done by a targeted population. In other words, different anthropometry (or manikin shape/size) needs to be tested in 3D. The term used to measure anthropometry is 'percentile'. A percentile is the value of a variable below which, a certain percent of observations fall. For example, the 20th percentile for a given anthropometry variable is the value below which 20% of the measurements may be found.

When using a DHM tool, there are different scenarios that might be adopted. In some cases, a specific anthropometry based on one single subject might be needed. Designing an F1 car for a specific driver is an example. In other cases, analysis might be done using only a 5th percentile manikin female and a 95th percentile male created from different anthropometry databases such as ANSUR, CAESAR, NHANES, for the US population or any other database from countries around the world. This is the case where multiple people need to be able to do a task. Think of a workplace on a car assembly line or designing a car that will be driven by a large number of people and sometimes, in many countries.

2.3.5 Produce a Report

Finally, a report is produced based on either milestone or specific time rate. The report can include ergonomic analysis results as well as vision windows, environment screen captures, manikin information and anthropometry, and so forth.

2.4 VERY SLOW DHM ADOPTION

Based on Chaffin (2009):

- Organizational structures often do not recognize the need for ergonomics and human factors early in the design process.
- Academic organizations are not providing many engineers and designers with even elementary human factors and ergo training.

Today, there is a gap between DHM research and end-user need. Many organisations do not recognize benefits that could be gained using DHM, considering model introduction timing, turnaround of those models and development costs that continue to escalate. The transferring of 'hands on' experience into virtual methods of analysis and the cost to initiate a program still create roadblocks for improvement. For many users, the connections are not made very quickly, and once understood, then identifying how to ask for improvements to the DHM software or requesting a study from a research facility seems overwhelming. Most users are required to use the software chosen by the organisation, which is not necessarily the latest technology available. Moving forward, for an organisation to reinvest in second or third software on top of the one chosen by the company, or to completely make a leap of faith to additional software, is very cost prohibitive. So, the user then begins to use and do what they can with the risk of missing opportunities to help the manufacturing processes.

Software companies look to delivering a DHM product to a company that is broad based but cost competitive. In many requests, the research is just unusable in a virtual environment. However, when it is available, more times than not, it is locked in a proprietary contract with a research facility. Many software companies are now held to delivering a partial product to their customers at no choice of their own.

Academic institutes and research facilities have many restrictions as well; they face proprietary agreements with software companies as well as manufacturing/design organisations. They are challenged in how research requests are made: does the request meet actual problems that exist as well as potential needs that would meet a virtual use? Would the grant or funding cover the expenses to meet either need or the research manage the added expenditures themselves? Finally, the research institute may or may not hold a high standard for DHM or decide to develop yet another DHM they themselves use. This creates opportunities for growth in the field, but locks valuable information away yet again from the user trying to improve tomorrow's jobs for those employees.

2.5   CONCLUSION

In conclusion, through an evaluation process, the user/engineer will look at how the DHM can simulate the interactions between the human being and his environment. However, when designing the part, product or workspace, the more powerful aspect is to consider how the design will create the interaction and what the flexibility will be. This thought is different from just reacting to an environment around the employee with changes, as they are needed. Typically, we expect the human to adapt to the product or workspace, but how does the design evolve as newer parts or products are introduced, without creating new risks for the employee?

Reducing the gap between the DHM research and the end users' needs is crucial; the cycle must be broken and the sharing of information needs improvements. There is a great opportunity for growth in the field of DHM at a faster rate of speed, but it will require efforts and actors negotiations for the betterment of DHM. All DHM stakeholders (users, developers and researchers) must collaborate in order for DHM to evolve.

ACRONYMS

CAD: Computer-Aided Design
CAM: Computer-Aided Manufacturing
DHM: Digital Human Modelling
IEA: International Ergonomics Association
OEM: Original Equipment Maker
PLM: Product Lifecycle Management
SAE: Society of Automotive Engineers
SgRP: Seating Reference Point
SMB: Small- and Medium-Size Business
VR: Virtual Reality

REFERENCES

Bubb, H. and Fritzsche, F., 2009. A scientific perspective of digital human models: Past, present and future. In V. Duffy (ed.), *Handbook of Digital Human Modelling, Research for Applied Ergonomics and Human Factors Engineering*, Boca Raton, FL: Taylor & Francis.

Chaffin, D. B., 2009. Some requirements and fundamental issues in digital human modelling. In V. Duffy (ed.), *Handbook of Digital Human Modelling, Research for Applied Ergonomics and Human Factors Engineering*, Boca Raton, FL: Taylor & Francis.

Chaffin, D. B., 2001. *Digital Human Modelling for Vehicle and Workplace Design*. Warrendale, PA: Society of Automotive Engineers.

Oudenhuijzen, A., Zehner, G. and Hudson, J. A., 2009. Verification and validation of human modelling systems. In V. Duffy (ed.), *Handbook of Digital Human Modelling, Research for Applied Ergonomics and Human Factors Engineering*, Boca Raton, FL: Taylor & Francis.

Ryan, P. W. and Springer, W. E., 1969. *Cockpit Geometry Evaluation Final Report*, Vol. V, JANAIR Report 69105, Washington, D.C: Office of Naval Research.

Reed, M. P., Faraway, J., Chaffin D. B. and Martin, B. J., 2006. The HUMOSIM Ergonomics Framework: A New Approach to Digital Human Simulation for Ergonomic Analysis, presented at the Digital Human Modelling for Design and Engineering Conference, Lyon, France.

3 Are You Sitting Comfortably? A Guide to Occupant Packaging in Automotive Design

Paul Herriotts and Paul Johnson
Jaguar Land Rover, UK

CONTENTS

In this chapter, the importance of considering ergonomics in vehicle design is discussed. The authors then describe a process that they have employed successfully in the premium automotive industry to package a range of occupants within a variety of vehicle types.

The aim of this chapter is to give the reader an understanding of the topic of occupant packaging in automotive design. However, it is not within the scope of this chapter to provide a detailed manual of vehicle occupant packaging; for the reader seeking such information, relevant Society of Automotive Engineers (SAE) papers and appropriate references are supplied.

3.1 INTRODUCTION

Let us start with some definitions. So, what do we mean when we use the term 'occupant packaging'? This term simply refers to the design of a vehicle around a specified range of drivers and passengers.

Packaging is the term that is used in the automotive industry to describe the harmonious placement of various components and systems in the vehicle space/architecture. What are these components and systems? Here we are thinking of such major systems and components as: powertrain (engine and gearbox); chassis (to include suspension, pedals and steering); electrical (instruments, electronic control unit (ECU), harness) and body ('body in white', interior and exterior trim and climate control systems). In addition to the major parts, there are hundreds or even thousands of minor components to be included (such as individual switches, relays, fuses, etc.).

Furthermore, occupant packaging is concerned not only with fitting these parts together, but doing so in a way in which they interact harmoniously (known as system and component compatibility). But for those of us working in occupant packaging, the key 'components' are of course the occupants, that is, the driver and passengers.

While we are looking at definitions, we should also define the term 'ergonomics' when used in an automotive design context.

Ergonomics or human factors engineering can simply be defined as the science of designing for people. By this, we mean applying knowledge of human characteristics and capabilities to the design of a vehicle. Gathering knowledge about people involves a multi-disciplinary approach with emphasis on anthropometry, biomechanics, psychology, statistics, and so forth. It is therefore a specialised role undertaken by those with formalised ergonomics training. The automotive ergonomist will interact with a range of automotive specialists: designers (exterior and interior) engineers from vehicle packaging, chassis, body, powertrain, electrical, and so forth. In addition to these technical designers and engineers who are responsible for developing vehicle geometry, the automotive ergonomist will also interact with those automotive professionals who are defining the vehicle segment versus competitors (product planning, market research, brand) as well as those people managing the development process such as project managers.

Each of these disciplines has an important role in the vehicle development process, but none more so than the automotive ergonomist who must ensure that the best possible fit between the vehicle and its occupants is achieved, with as large a range of people as possible accommodated comfortably. That is, the full range of drivers and occupants within the intended population are considered and catered for in terms of comfort, ease of use and also of safety.

As with any introduction to ergonomics, it is very important to note that the central philosophy of the discipline is to fit the product to the user, and not the user to the product. Thus, in the automotive setting, we do not want the driver to adapt to a compromised, poor design, but instead we must design a vehicle to meet his or her needs and capabilities.

3.2 THE AIMS OF OCCUPANT PACKAGING

Let us start with the driver in a logical progression of interaction with the vehicle. Firstly, as he/she approaches the vehicle, he/she must be able to get in easily. So, ease of entry and exit is our first important consideration. Secondly, having easily entered the vehicle, the driver must be able to achieve a comfortable driving position (see Figure 3.1). When we talk about a comfortable driving position, we are thinking not only of fitting in the vehicle in the car showroom, but also using the vehicle in comfort for long journeys over a sustained period of time. The occupant packaging aims therefore are to ensure that a wide range of drivers can achieve such a comfortable driving position; that they can easily reach and operate the primary and secondary controls in comfort, that they can easily see the in-car displays from their chosen driving position, and that they have good all-round vision out of the vehicle. These are the basic aims for packaging our driver.

Now let's consider the passengers. Once again, they must be able to get in and out of the vehicle easily. This is not as simple a task as it sounds when the variety of passengers will include a newborn infant in a child seat being placed in the vehicle by a parent, a young child getting in unassisted on a busy road, an older lady with arthritis or an obese man with significant mobility impairment, and so forth. As described above, these people expect the vehicle to be designed to fit their needs and abilities; why should they adapt themselves to a compromised

FIGURE 3.1   The aims of occupant packaging: a comfortable driving position.

design? From a commercial perspective, it is vital to remember that they can vote with their feet and seek out a competitor vehicle that does provide what they want.

Once the passengers have easily entered the vehicle, they must have appropriate room/space within the vehicle and be able to sit in a comfortable posture. As with the driver, they must be comfortable over a period of time, so we should design the vehicle to allow them to change posture and to have enough space to do so. As with the driver, the passengers must also have an acceptable level of vision out of the vehicle.

The ambience of the cabin is another very important factor to consider and it should feel an inviting and comfortable environment to both drivers and passengers. The amount of light entering the cabin can be critical to this feeling. This is particularly a focus in premium vehicles; how occupant packaging can be used as a brand or vehicle differentiator from more mundane competition is now discussed.

3.2.1 OCCUPANT PACKAGING AS A BRAND DIFFERENTIATOR

In addition to the basic occupant package requirements, occupant packaging may be used as a vehicle or brand differentiator particularly within the premium segment of the automotive industry.

Let's examine this concept in more detail. It is possible for example to design in a feeling of vehicle sportiness as typified by a Jaguar sports car (see Figure 3.2), so that when the occupants (driver and passengers) approach the car, get in, get comfortable within the cabin and look out over the bonnet, their posture, environment and vision affords them a feeling of sportiness even before they start the car and move off. There is a feeling of sitting 'in' the vehicle rather than 'on' the vehicle. The car has been made to feel special through occupant packaging.

Similarly, we may wish to give the drivers the feeling of being in command of their vehicle, as typified by a Land Rover (see Figure 3.3). So once again, as drivers approach the car, get in, get comfortable within the cabin and look out over the bonnet, their posture, environment and vision affords them a supreme feeling of command and being in charge or control of the car and a good sense of the external

FIGURE 3.2   Jaguar C-X16 concept.

FIGURE 3.3   Land Rover DC100 concepts.

environment. This feeling of command is again present even before they start their Land Rover and move off.

It is interesting to note that while occupant packaging has contributed to these vehicles feeling as they do, the end result of the feeling of sportiness or being in command, has been achieved using the same development process, but using a very different occupant packaging 'formula'.

The formula for sportiness includes positioning the driver in a low-slung posture, positioned close to the road. The cabin is made to feel intimate, achieved with a number of touch points between the driver and the vehicle and by appropriate positioning of the cabin architecture (e.g., the upper canopy) in relation to the driver and by the driver being positioned close to the front passenger. The relationship between the position of the driver and the vehicle dash/fascia is optimised to ensure the driver is 'tucked into' the car and feels intimate with it. The size of the steering wheel and position of the gear lever relative to the driver and steering wheel is also considered. Other factors to consider include the height of the waistline of the car relative to the driver's hips and eyes, and the form of the bonnet/wings to allow the driver to see a pronounced feature above the front road wheels which help in positioning the car at speed. The layout of the pedals is optimised to allow the driver to heel and toe. The term 'heel and toe' is used to describe a driving technique where the driver depresses the accelerator and brake at the same time with his/her right foot, so maintaining engine speed during cornering and therefore faster driving.

The formula used to develop the feeling of command includes positioning the driver in an upright posture, high above the ground. The cabin is made to feel open and airy, achieved by thoughtful positioning of the cabin architecture (e.g., the upper canopy) in relation to the driver. Appropriate space is given to the driver in relation to the vehicle architecture and he/she is positioned away from the front passenger. Superior vision out of the vehicle is critical, so the driver is positioned in an outboard location, close to the edge of the vehicle and the height of the bonnet, dash/fascia and the vehicle waist in relation to the driver's eyes is optimised, as is the design of the A-pillars.

So clearly, if we think of premium cars such as Jaguars, Range Rovers, Aston Martins or Rolls-Royces, it is clear that the cars feel very special and very different from what could be termed more mainstream offerings. Part of this special feeling is generated by the cabin ambience and hence occupant packaging, as well as, of course, vehicle styling, materials, perceived quality, and so forth.

So, occupant packaging has an important role to play, not only in the basics of accommodating the range of drivers and passengers, but as a vehicle and brand differentiator.

3.2.2 THE ROLE OF OCCUPANT PACKAGING IN CAR DESIGN

The description of occupant packaging given above may sound straightforward. Surely, every vehicle must comfortably accommodate the driver and passengers? Well, at this stage of the chapter it is interesting to consider why occupant packaging is in fact such a difficult and challenging task. After all, we are simply describing the fundamentals of automotive ergonomics, and surely everyone must recognise their importance? Well, not quite. Generally in the automotive industry, we see a preoccupation with design (aesthetics). In a competitive marketplace, exterior design in particular is a clear differentiator between vehicles and is of significant importance in the buying decision. As a consequence, the aesthetic aspect of vehicle design often takes precedence over the less obvious aspect of vehicle ergonomics. It is of great importance then that ergonomics input to vehicle design takes place from day one of the development process and continues throughout the process. Only by engaging with the design (and engineering) teams will the ergonomist be able to influence the design to meet the ergonomics goals that have been set.

The automotive design process has historically been based on rival teams or individual designers with a strong element of competition. Clearly, each designer hopes that his or her design will be the chosen one: a single successful vehicle design can be career defining, so the pressure to generate a beautiful looking exterior is significant. Consequently, at the initial phase of the design process there is a strong temptation to focus on the aesthetics of the exterior design at the expense of functionality, as this may be viewed as an inhibitor to exterior design. There is also the feeling that the design can be subsequently modified to include the occupant packaging criteria. However, this philosophy of designing from the 'outside in' is significantly flawed, as it presents severe challenges to the vehicle development team, often with resultant compromises. Successful design is based on meeting a set objective within clearly defined constraints. Compromise does and will occur, but it must be limited and understood. This will not be possible if ergonomics objectives are not considered from day one of the design process. The concept of designing from the 'inside out' is therefore recommended, with a set of agreed physical 'hard points' that allow good occupant packaging to be defined, around which the designers base their initial designs. Examples of design renderings that are based around package hard points and provide the driver with good exterior vision are shown in Figures 3.4 and 3.5.

We should also understand that in addition to exterior aesthetics, occupant packaging is only one of many challenges facing automotive engineers and designers and, as such, may sometimes be compromised. Other considerations during the vehicle

FIGURE 3.4   An initial design rendering respecting occupant package hard points.

FIGURE 3.5   An initial design rendering giving the driver good exterior vision.

development process include legal requirements, quality, serviceability, aerodynamics, weight, ease of manufacturing, vehicle dynamics, performance and economy, refinement, durability and reliability, safety and security, as well, of course, as the important consideration of cost.

So, while the authors consider the driver and passengers to be of supreme importance in vehicle design and engineering, their counterparts working in other areas of automotive engineering will have a different emphasis and consequently fight for their own particular attributes. This is clearly demonstrated by the fact that every vehicle on the road today is a compromise in some way, presenting a mix of vehicle attributes. It is of importance then that the process described later in this chapter ensures that such compromises are made based in an informed manner, with known ergonomics consequences.

Further challenges arise from the engineering community. As the physical components of a vehicle are well defined from a geometric perspective, they are well understood by the engineer. The human occupant is less well defined geometrically, so it is always tempting for the engineer to assume that any compromise can be made by the component that is adaptable, that is, the human occupant. This is one of the many pressures in occupant packaging and why the automotive ergonomist needs robust data to set targets for occupant accommodation.

The next section describes a process that starts in a relatively coarse, broad-brush manner, and then is refined and polished as the design matures.

3.3   THE OCCUPANT PACKAGE DEVELOPMENT PROCESS

In the development of any vehicle, the level of detail/reality increases from an initial 2D design sketch or rendering, to virtual 3D images to physical prototypes. The process reflects this, starting with 2D geometry and basic dimensions (SAE J1100 and internal company guidelines), laying out the basic architecture of the vehicle (see Figure 3.6). It then moves on to computer-aided design (CAD) modelling of a virtual vehicle, driver and passengers (using CAD tools such as CATIA and the RAMSIS human digital modelling software), then on to physical assessments of early proto-type vehicles (known as 'bucks') using physical assessors (people). Each of these stages is described in detail in the following sections.

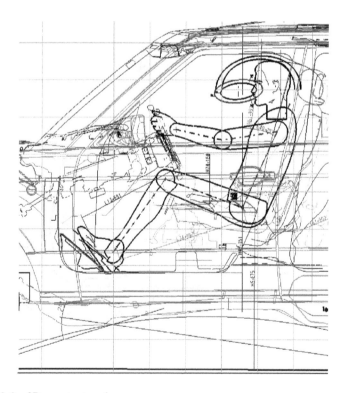

FIGURE 3.6   2D occupant package.

3.3.1 STEP ONE: EARLY DESIGN DEVELOPMENT USING SAE STANDARDS AND INTERNAL COMPANY GUIDELINES

In simple terms, this first step in the process aims to set the driver's posture and then to lay out the initial vehicle architecture around him/her.

In doing so, various SAE standards are referred to. The Society of Automotive Engineers International (SAE) is a standards development organisation for engineering professionals in the aerospace, automotive, and commercial vehicle industries. It publishes over 1,600 technical standards relating to the design of passenger cars; included in these are a number of ergonomics standards that are critical to vehicle occupant packaging.

They provide recommended practice for laying out the initial package, however care must be taken in their use, and any occupant package must be verified and refined through human evaluations with physical architecture. It is also important to note that SAE standards may be more applicable to a US population, whereas vehicles developed for worldwide markets will require detailed data relating to those markets to refine the occupant package appropriately.

Before the occupant locating process is described, it is important to first consider the driver's hip point (H-point). The H-point describes a theoretical intersection of the occupant's thigh and torso lines (see Figure 3.7).

Using the 3D H-point machine (SAE J4002) as shown in Figure 3.8, the location of the H-point relative to the physical seat can be determined. This can then be used to position the H-point of a virtual 2D H-point template (see Figure 3.7) in relation to a CAD seat. Thus, the H-point can be used to relate a physical seat to virtual geometry.

Whereas on a fixed (non-adjustable) seat there is only one H-point position, on an adjustable seat (e.g., the driver's) the H-point can be located in a number of positions. The extremes of these can be mapped and described as the seat movement envelope. In order to have one point of reference for occupant packaging, the manufacturer will create a unique design H-point known as the seating reference point (SgRP). This is the reference point used to position the SAE 2D template.

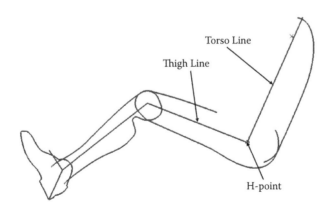

FIGURE 3.7 H-point template.

FIGURE 3.8   SAE H-point machine.

The SgRP is a fundamental reference point for defining and describing both the occupant package and vehicle dimensions (see Figure 3.9). Many of the occupant-related factors and legal requirements are quoted in relation to the SgRP: the occupant packaging engineer and automotive designer will therefore need to have a good understanding of this. In addition, the SgRP enables correlation between the virtual and physical environments, providing a consistent method for the comparison of vehicles (internal and competitors).

Having set up the driver's template, a variety of CAD tools and recommendations are used to establish the space around the occupant, locate the primary and secondary controls, and define direct and indirect fields of view.

Examples of the recommendations used to establish the occupant package include the following:

Head contours as defined in SAE J1052 are used to help establish the upper cabin architecture of the vehicle, so ensuring the driver has enough space around the head.

The driver's hand control locations as defined in SAE J287 are used to help establish the location of the primary and secondary controls, so ensuring that they are in reach by the driver.

The driver's eye locations can be represented using 'eyellipses' as defined in SAE J941; these are used to help establish the vehicle architecture, so ensuring that the driver has adequate vision.

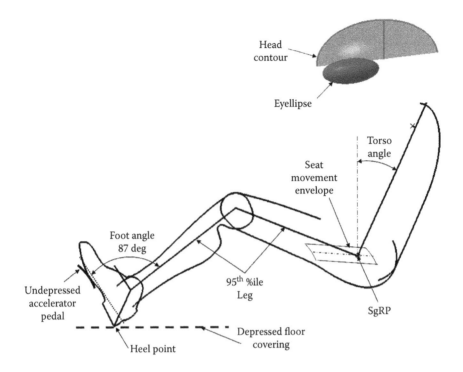

FIGURE 3.9 Driver H-point template.

Having established the driver's package, the rear passenger templates can now be located. This enables their posture, location in relation to the driver and the space around them to be defined.

Internal company guidelines are used in conjunction with SAE direction. Such company guidelines are based on experience of developing vehicles in the segment, of acceptable and preferable dimensions, and may also refer to the benchmarking of competitor vehicles, that is, to be competitive with the existing set of competitor vehicles, adopt the mean dimensions relating to occupant space.

At the end of this first step in the occupant package development process, we would expect to have an initial occupant package that acts as a good starting point. By that, we mean that the package has no major flaws. However, it is very far from being a completed occupant package and requires significant further development and refinement to successfully meet the needs of the user population.

3.3.2 STEP TWO: VIRTUAL DEVELOPMENT AND ASSESSMENT OF THE OCCUPANT PACKAGE

Computer-aided design (CAD) tools with digital human models are available to be used in the ergonomics design process. These human models can be configured to represent people of various shapes and sizes in many populations, and so represent the intended user group for any vehicle. Such specialist software packages include

RAMSIS, Jack and SAMMIE, as well as human models available within for example CATIA (Human Builder) or ALIAS CAD packages.

The purpose of these CAD tools is to predict the interaction of people with a physical environment. As such, any software chosen should first have been validated to ascertain if for example, the predicted postures, reach and vision do indeed match those experienced by human occupants in a physical environment. Clearly, human behaviour is complex and so it is difficult to model. Many engineers assume that these digital human models represent the answer to all their ergonomics queries, and that they can successfully package a vehicle using CAD tools with great confidence in the output. Digital human modelling does bring many benefits to the design process, but should be used as a crude filter to remove the more obvious occupant packaging issues. User trials with representatives from the user population with a representative buck will highlight issues that are not evident using digital human modelling, including long-term comfort issues, effects of fatigue, and a range of subtle issues such as product acceptance based on past experience. The main advantages of digital human modelling are that the occupant package can be developed and assessed early in the vehicle programme without the high costs associated with designing, building and assessing a physical buck. In addition, by using a digital human modelling tool to develop the initial occupant package, when a physical buck is subsequently built, it represents a design that is already refined to a certain extent; most of the major issues should have been ironed out using virtual assessments and so the basics of the occupant package should be in place.

Comparisons between the available CAD tools are made in the literature and it is beyond the scope of this chapter to provide a detailed description of each software package or an overview of their relative merits. For the purposes of vehicle development at Jaguar Land Rover, the authors use RAMSIS software, described in detail in the next section.

3.3.2.1   A Note on Package Drawings

The initial vehicle package is developed as a set of 2D sections, which are cut at various critical planes through the vehicle. These are produced by a packaging engineer to show the basic geometric layout of the vehicle and so to convey the logic behind the vehicle architecture. The package drawings are shown in three views: from the side, in plan view and from the front of the vehicle. An example package drawing is shown in Figure 3.10. As may be evident, the packaging engineer has included 2D representations of the occupants within the vehicle, as well as the key vehicle components. The drawings also show the three-dimensional grid reference system (X, Y and Z coordinates) that is used to relate the vehicle dimensionally. A variety of ground lines are also shown to represent the vehicle attitude when loaded to different conditions (i.e., the vehicle will sit differently on the road when it is empty to when it is loaded with luggage and occupants).

Critical vehicle dimensions are included; all dimensions are measured in millimetres using a system described in SAE J1100 with the prefixes L, W, H, A, D and V denoting dimensions of length, width, height, angle, diameter and volume. Some package drawings may also show a competitor vehicle's dimensions for benchmarking purposes.

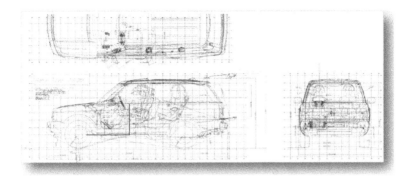

FIGURE 3.10   A package drawing of a Range Rover.

When examining a package drawing, it is possible for an experienced engineer to get a feeling for the relationship between the driver and his/her surrounding vehicle architecture. However, it must be remembered that the 2D H-point template represented in the drawing is an SAE derived manikin, with little representation of the true range of occupant sizes and postures. As such, it is useful for comparative purposes when, for example, comparing one vehicle to another, but is of less use in assessing where in reality the occupant will actually position him/herself.

However package drawings are the accepted means by which the vehicle geometry is often collated and communicated within a vehicle manufacturer, and as such the occupant packaging professional must be able to 'read' a package drawing and understand the issues it conveys.

3.3.2.2   Virtual Tools: RAMSIS Human Digital Modelling Software

The RAMSIS human digital modelling software allows a range of human occupants to be easily modelled (see Figure 3.11). These models display varied body dimensions which can be set by the RAMSIS user, so enabling, for example, a short stature,

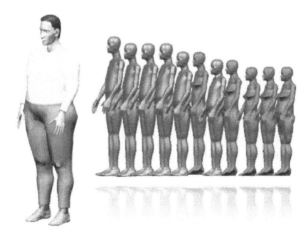

FIGURE 3.11   RAMSIS digital human manikins.

FIGURE 3.12   A male RAMSIS manikin in a representative driving position.

young woman with a long back to be modelled, or an obese older, taller man with short arms. In addition, data can be selected from populations around the world, to reflect the intended markets for the vehicle. As populations are not static but change with time, RAMSIS software can also predict the sizes of people in the future, so that human models can be built of a driver, say, 10 years into the future. This, of course, is important when a vehicle model may be on the market for a sustained period of time.

Having modelled human occupants, what does the user do with these manikins? The software enables them to be positioned in a driving position within a CAD geometry model of the vehicle under development (see Figure 3.12). The ability to get comfortable, to reach the primary and secondary controls and to see these controls and the in-car displays, as well as vision out of the vehicle can be assessed.

The positioning of these RAMSIS digital human manikins is based on knowledge acquired of where humans sit in vehicles, and so the RAMSIS digital human postures that result are a good indication of where people will really sit in the vehicle that is built to the geometry being assessed.

Using a specific process with specified tasks and manikin sizes, the user can undertake repeated assessments of the initial occupant package and develop it to ensure that the basic aims of occupant packaging are met, that is, that the intended range of drivers and passengers are accommodated within the vehicle package.

The authors have validated RAMSIS output using physical buck studies with a range of drivers and passengers, and found RAMSIS to be a good indicator of occupant positioning. As such, recommendations made to amend/develop vehicle geometry based on RAMSIS output are of worth, and this early refinement of the occupant package saves a significant amount of development time and money.

At the end of this stage of the occupant package development process, we would expect to have an occupant package that theoretically begins to meet the needs of its intended user population, but still requires further development. We should now have reached a point where we have enough confidence in our occupant package design to commit to spending significant resources (up to around $200,000) in designing and building an accurate physical model to represent it. However, one further step may be added and is described in the next section.

3.3.3   STEP THREE: THE VIRTUAL REALITY CAVE

Many automotive companies are now using virtual reality cave facilities to aid the vehicle development process. This immersive environment is simply a space with three walls and the ceiling acting as screens onto which high-resolution images are projected. Wearing '3D glasses', as in the cinema, then allows the user to easily understand and experience a vehicle's geometry in 3D (see Figures 3.13 and 3.14).

This is a very powerful development tool as the visual experience from a virtual driving position in the cave correlates well with that experienced in a real vehicle. This gives the vehicle engineer the ability to assess vehicle geometry before a physical model has been built. Moreover, it is possible to rapidly and easily assess alternative designs or competitor vehicles for comparison purposes.

However, it should be noted that this is merely a further step in the development process albeit an important one; at the time of writing, the cave is no substitute for assessing a physical vehicle. For example, ease of entry and exit still require physical assessment.

The pioneering VR cave employed by the authors in their work at JLR's Gaydon research and development facility, has also proven to act as a very useful communication tool; it can be used to quickly and clearly show an embryonic vehicle to the project team. In doing so, it brings to life a vehicle that exists only as 3D geometry in a virtual world and so helps the packaging engineer in getting support and buy-in of his proposals.

FIGURE 3.13   VR cave models being manipulated in the cave space by the user.

FIGURE 3.14   Range Rover Evoque development in the JLR cave at Gaydon's R and D facility: exterior vision optimised.

At this stage of the process, the vehicle package is becoming increasingly refined and aspects of the vehicle related to interior and exterior vision are now optimised. This would, for example, include the profile of the bonnet.

The next stage describes the modelling process.

3.3.4 STEP FOUR: PHYSICAL MODELLING USING BUCKS AND USER TRIALS WITH CUSTOMER REPRESENTATIVES

This stage of the occupant packaging process involves physically modelling both the vehicle and the intended occupants (target customers). Let us look at each in turn.

How is the vehicle modelled? A 'buck' is a physical model of the vehicle, accurate to within a few millimetres. Aspects of the design that are critical to occupant packaging are included in as accurate a manner as possible, while aspects that are non-critical to occupant packaging are expressed only in a crude form (materials are often not representative of those in a car, but the buck is nevertheless good enough to give a feeling of what the final vehicle would be like). This is an expensive tool that takes significant resources to produce. Before a buck is evaluated, the SAE H-point machine must be used to ensure the seat envelope is representative of the intended geometry.

Bucks can be static or dynamic. Static bucks are assessed in a fixed position within a laboratory, and are initially used to gain data of customer satisfaction of such attributes as ease of entry and exit, comfort of the driving position, and so forth (see Figures 3.15 and 3.16). Dynamic bucks/early dynamic prototype vehicles can be driven, and so assessed dynamically (see Figures 3.17 and 3.18). They tend to have a greater degree of representation and so they are extremely expensive and time consuming to produce. However, they are vital in the development and assessment of

FIGURE 3.15 Taller passenger in a simple buck designed to assess entry to and exit from the rear of a Range Rover.

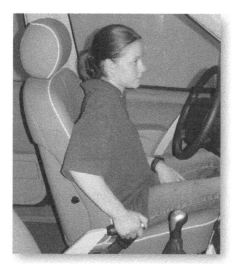

FIGURE 3.16 A shorter female driver assessing the driving position and control placement in a static buck in laboratory conditions.

those vehicle attributes that are best assessed dynamically such as the design of door mirrors, which must be assessed while parking.

Let us now consider 'modelling' the intended drivers and passengers. Section 3.3.2.2 described modelling people using the RAMSIS digital human modelling software. But later in the process, when we have physical models of vehicles to

FIGURE 3.17 A dynamic buck with a competitor production vehicle for comparison purposes on the JLR Gaydon test track facility.

FIGURE 3.18 A prototype vehicle (wearing camouflage to disguise the exterior styling) being dynamically tested.

assess, we must model the drivers and passengers who will use the vehicles by selecting people who represent as closely as possible the intended customer base around the world. The human assessors must be carefully chosen and an appropriate number must be used in assessments.

At JLR, we have a number of internal panels of employees who act as customer representatives, as well as carrying out external customer clinics with potential customers. Members of these internal panels participate in buck and vehicle evaluations (see Figure 3.19). They are measured and their anthropometry recorded; when members of these panels are selected to take part in buck assessments, they are chosen based on their anthropometry to ensure that the extremes of the intended user population are covered.

It is beyond the scope of this chapter to discuss population sampling in detail, and the reader is referred to the further reading section.

FIGURE 3.19   Members of JLR women's panel.

3.3.4.1   A Note on the Diversity of Occupants

One reason why occupant packaging is so difficult within a challenging automotive design and engineering environment was discussed earlier in the chapter (many designers and engineers fighting for their own attributes). But there is a bigger picture to consider when asking why occupant packaging is so difficult, and that considers human diversity and the vast range of people that could be catered for. Human diversity is considerable, and this is particularly evident when we are faced with the task of designing a single product for a global population.

JLR vehicles are sold in over 170 countries around the world. Within each of the countries, the vehicles must be suitable for a wide range of people and one vehicle model must be suitable for all countries, that is, one size fits all. Thus, all these populations must be considered when vehicles are designed and developed. In recent years, JLR and the vehicle industry in general has faced the challenge of designing for emerging markets. The countries of Brazil, Russia, India and China are known by the acronym, BRIC, and are grouped together as they are considered to be at a similar stage of recently advanced economic development. For a world vehicle design to be successful, it must be designed to meet the needs of these BRIC populations; knowledge of these populations is therefore vital.

Clearly the challenge of being user centred and designing to accommodate a large population is difficult; we must consider the physical human differences that result from age, gender, nationality, ethnicity, changes over time, lifestyle, and so forth, when designing our vehicles. How do we measure these differences? We focus on anthropometric data. It is time for another definition: simply put, anthropometry is the measurement of body dimensions. Anthropometry has been established for many years and has been critical in the understanding of human variation and has

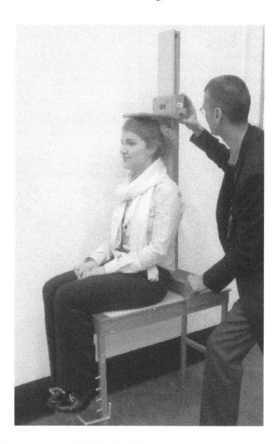

FIGURE 3.20   Measurement of 'sitting height'.

provided critical guidance to ergonomics and design. How is such human variability expressed? Anthropometric data is used to describe the central tendency or mean of a population as well as the distribution of the data in the population, with 5th and 95th percentile figures generally quoted. There are over 200 standardised dimensions that have been taken from various populations around the world (see an example in Figure 3.20).

The ergonomist must ensure that any anthropometric data used to guide a design is as recent as possible. Within a population any changes in diet, healthcare, lifestyle, age profile, and so forth, may lead to rapid change in the anthropometric data of a population.

Static anthropometry refers to the measurement of body dimensions taken with the body held in a number of standardised, defined static postures. Stature or sitting height would be such a dimension. So, by their very nature, anthropometric data are gathered from people in defined postures. However, these defined postures bear little relation to the various postures adopted by drivers, for example, in the complex dynamic tasks of getting in and out of a vehicle or of driving. Dynamic anthropometric data collected under 'functional' conditions do exist, but they generally relate to simple tasks such as reach.

An excellent source of general anthropometric data can be found in Adultdata (Peebles and Norris, 1998), but when referring to ergonomics and consumer product design, even the authors of this anthropometry 'bible', note: 'a programme of testing and evaluation—involving technical, user and simulation testing—is always necessary'.

3.3.4.2 User Trials

So, participants who have been chosen based on their anthropometry (to ensure that the extremes of the intended user population are represented), evaluate a buck or a working prototype on a one-to-one basis, in a controlled laboratory or test track environment. The purpose of such a user trial is to elicit information from a group of participants relating to user satisfaction, perception and expectations of a prototype design.

The automotive ergonomist has many tools at his disposal: interviews, questionnaires and observation methods are used in this stage. Interviews are flexible and a useful, good tool for collecting user perception data. Participants are interviewed in a semi-structured manner on a one-to-one basis. Similarly, questionnaires offer a good means of rapidly collecting data from participants, with a focus in this process on user satisfaction. In our process, these are administered by the experimenter on a one-to-one basis. They are a good tool to evaluate design concepts and to probe user satisfaction, being flexible, allowing easy data analysis, and being easy to administer, and so forth.

Finally, observation is undertaken of participants undergoing a complex activity such as entry or exit. Detailed task performance can be recorded and analysed, but the very act of recording may change participant behaviour and is time consuming to analyse.

It is important to undertake these assessments with participants put at ease; a permissive environment must be provided where participants feel free to criticise a design and to make negative as well as positive statements without being judged or receiving adverse comments from the experimenter. As such, the trials are conducted on a one-to-one basis, in a relaxed informal manner.

Feedback from the trials is then interpreted into design recommendations by the ergonomist. For example, questionnaire feedback from participants relating to difficulty in entering a vehicle would trigger an observational study where it might be evident that the sill profile on the buck was causing a contact point with the leg. A local change in sill profile would address the difficulty being experienced. The ergonomist would therefore recommend a new sill profile be adopted, and would specify its geometry to the relevant engineer. If these recommendations are adopted or more typically partly adopted, the buck is updated to reflect this revised geometry. The buck is then retested with the aim of confirming if the changes have been effective in improving the occupant package. This is an iterative process, with many revisions and testing iterations being typical.

Having refined the occupant package, the finalised design will meet the needs of the intended user population. It will be easy to get in and out of the vehicle, a comfortable driving position will be found with controls within reach and falling to hand, and interior and exterior vision will be good.

Using people in the latter stages of this development process is a powerful tool. Because the package has been refined using so-called participatory design principles, we can have great confidence that our customers will find that it meets their needs well.

3.3.5   Step 5: Validation of the Finalised Vehicle; Feedback from Customers Who Buy and Use the Finalised Vehicle

J.D. Power conducts multiple annual surveys of the automotive industry in the US as well as in other countries. The APEAL survey reflects consumers' attitudes towards a vehicle's attributes. JD Power industry data and company-led surveys provide the engineer with customer feedback of their own brands' models as well as that of competitor vehicles. This data is invaluable in that it allows the engineer to understand how the vehicle geometry impacts on the customer's perception of, for example, vision, roominess, ease of entry and exit.

This information will hopefully validate the findings of the in-house development process and will also inform future vehicle development.

3.4   SUMMARY

In this chapter the reader has been provided with an understanding of the topic of occupant packaging through a discussion of the importance of considering ergonomics in the vehicle design process and a description of the process employed successfully in the premium automotive industry to package a range of occupants within a range of vehicle types.

The authors' desire was to convey their passion for occupant packaging and in particular for being user centred: 'design from the inside out' is the mantra.

The reader seeking more detailed information is referred to the section 'References and Further Reading', below.

The authors of this chapter, Dr Paul Herriotts and Paul Johnson, hold the positions of 'Technical Specialists' with the premium car manufacturer, Jaguar Land Rover.

Paul Herriotts is the ergonomics specialist, while Paul Johnson is the specialist in cockpit and cabin packaging. They have provided technical input to numerous iconic and successful British cars, ranging from Mini to Rolls-Royce Phantom, including recent Jaguar, Land Rover and Range Rover vehicles.

As technical specialists at JLR, the authors are responsible for occupant packaging of vehicles that are sold in over 170 countries around the world.

ACKNOWLEDGMENT

Many thanks to Andy Wheel for providing design renderings.

REFERENCES AND FURTHER READING

Bhise, V., 2012. *Ergonomics in the Automotive Design Process*. Boca Raton, FL: CRC Press, Taylor & Francis.

Macey, S. and Wardle, G., 2009. *H Point: The Fundamentals of Car Design and Packaging*. Culver City, CA: Design Studio Press.

Peebles, L. and Norris, B., 1998. *Adultdata: The Handbook of Adult Anthropometric and Strength Measurements*. London: Department of Trade and Industry.

Pheasant, S. T., 1984. *Anthropometrics: An Introduction for Schools and Colleges*. London: British Standards Institution.

Salvendy, G. (ed.), 2006. *Handbook of Human Factors and Ergonomics*, 3rd ed. New York: John Wiley and Sons.

SAE J287, 2009. Driver hand control reach. In *SAE 2009 Handbook*, Warrendale, PA: SAE.

SAE J826, 2009. Devices for use in defining and measuring vehicle seating accommodation. In *SAE 2009 Handbook*. Warrendale, PA: SAE.

SAE J941, 2009. Motor vehicle drivers' eye locations. In *SAE 2009 Handbook*. Warrendale, PA: SAE.

SAE J1050, 2009. Describing and measuring the driver's field of view. In *SAE 2009 Handbook*. Warrendale, PA: SAE.

SAE J1052, 2009. Motor vehicle driver and passenger head position. In *SAE 2009 Handbook*. Warrendale, PA: SAE.

SAE J1100, 2009. Motor vehicle dimensions. In *SAE 2009 Handbook*. Warrendale, PA: SAE.

SAE J1516, 2009. Accommodation tool reference point. In *SAE 2009 Handbook*. Warrendale, PA: SAE.

SAE J1517, 2009. Driver selected seat position.In *SAE 2009 Handbook*. Warrendale, PA: SAE.

SAE J4002, 2009. H-point machine (HPMII) specifications and procedures for H-point determination. In *SAE 2009 Handbook*. Warrendale, PA: SAE.

SAE J4004, 2009. Positioning the H-point design tool—Seating reference point and seat track length.In *SAE 2009 Handbook*. Warrendale, PA: SAE.

4 IVIS, ADAS, OODA
Joining the Loops

Nick Reed
Transport Research Laboratory, UK

CONTENTS

4.1 INTRODUCTION

Modern life places demands on our capabilities to multi-task. The basic drives to satisfy hunger, to sleep and to reproduce are supplemented by a variety of goals that may be directly related, indirectly related or independent of these motivations. In this chapter, the nature of driving as a sub-task within the framework of human goals shall be discussed.

4.1.1 THE DRIVING TASK

In the early twentieth century and soon after the emergence of the automobile, pioneering manufacturers, Daimler Benz, made plans to cope with expected future demand for automobiles, predicting that in a century, there would be one million cars on the road. This prediction was based on the assumption that the maximum number of chauffeurs who could be trained to drive was one million (Peppers and Rogers, 2008). The prediction was based on a false assumption about the difficulty of the driving task. The speed with which the car enabled movement of people, goods and information led to development pressure to improve the usability of the automobile. This resulted in vehicles that were relatively easy to drive and maintain compared to their forebears and so driving became viable to a much larger proportion of the population than Daimler Benz had envisaged. By 2000, there were in fact 600 million cars worldwide, with annual production of around 60 million.

However, the apparent ease with which driving can be accomplished is at odds with the true complexity of the task; McKnight and Adams (1970) estimated that driving is composed of more than fifteen hundred sub-tasks. Given this complexity, it is perhaps surprising that Stutts et al. (2005) found many drivers engaged in driving-unrelated tasks such as eating, smoking, reading or using a mobile phone in an observational study. Nevertheless, this willingness to engage in additional tasks when driving has consequences. In a large-scale naturalistic driving study, Dingus et al. (2006) reported that 78% of the observed crashes were associated with driver inattention. Using data from the same naturalistic driving study, Klauer et al. (2006) found that drivers were engaged in secondary tasks for 23.5% of their time when driving and that tasks involving moderate and complex manual/visual interactions were found to have a significantly higher association with safety critical events (i.e., crashes or near-crashes).

4.1.2 Mobile Phones and IVIS

A relatively common driving-unrelated task in which drivers choose to engage is the use of a mobile phone. Observational studies across Europe, Australia and the US have found that 1 to 6% of drivers are engaged in phone calls when driving (Breen, 2009). Telephone communication in cars began in the 1940s (Klemens, 2006) when car phones first became available. The service at the time was limited; the equipment cost more than the price of a new car, calls had to be routed via an operator, connections were frequently lost and as the service gained customers, demand quickly exceeded the capacity of the network so callers often had to wait for a line to become available. Although progress in mobile telephony was made in the latter half of the twentieth century, it was not until the advent of digital cellular networks in the 1990s combined with further miniaturisation of the technology and improved affordability that uptake became widespread. Figure 4.1 uses data from the World Bank to show how market penetration of mobile phones increased rapidly in Europe and the US through the 1990s and worldwide in the 2000s (World Bank, 2012) to the extent that mobile phones are near ubiquitous.

The use of a mobile phone enables a driver to remain contactable at all times but the digital cellular networks permit transmission of more than telephone calls. The exchange of data, combined with knowledge of vehicle position available from satellite systems and on-board computing power, creates a rich source of information for use within a vehicle and by a driver. Design guidance (e.g., Stevens et al., 2002; Stevens, 2008) has been developed for such in-vehicle information systems (IVIS). This guidance covers the presentation of information and the design of the interfaces that enable access and use of this information by a driver. The aim of the guidance is to maximise safety and usability of IVIS. However, Lee and Strayer (2004) highlight the usability paradox as an issue for the safety of driving whereby the greater the ease of use of a system, the greater the likely frequency of use of the system. Consequently, the cumulative time over which a driver is distracted may be greater

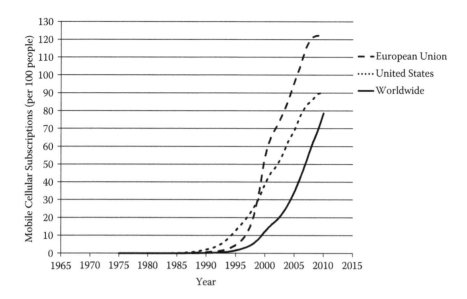

FIGURE 4.1 Market penetration of mobile phones over time in Europe, the US and Worldwide. (From World Bank (2012), Mobile Cellular Subscriptions (Per 100 People), Washington, DC. Available at: http://data.worldbank.org/indicator/IT.CEL.SETS.P2, Accessed: February 2012.)

for a system that has been well designed in terms of usability than a poorly designed system that is rarely used with a commensurate increase in risk.

Recognising the distraction risk posed by increasingly complex IVIS and nomadic devices, National Highway Traffic Safety Administration (NHTSA) (2012) published draft voluntary guidelines on driver distraction that list explicitly non-driving tasks that they consider 'interfere inherently with a driver's ability to safely control the vehicle'. These include the display of images or video not related to driving and manual text entry tasks. The guidelines state that devices should be designed to reduce the requirement for drivers to take long glances away from the roadway such that tasks can be completed using glances of less than two seconds and a cumulative time of twelve seconds with eyes away from the road. NHTSA's guidance in relation to cognitive distraction suggests that this is a far less significant component in accident risk. Conversely, Strayer, Watson and Drews (2011) report the results of numerous studies that show clear impairments to various aspects of driving performance by cognitive distraction.

Koch (2004) described human 'zombie' behaviours as being those where actions occur without the conscious mind being aware of the stimulus and physical response. An example of a zombie behaviour is thermoregulation—we are not consciously aware of the processes that maintain body temperature. Engström (2011) suggests that zombie behaviours, inflexible but efficient in responding in routine situations, may be a good representation of what happens in driver cognitive distraction. Zombie driving may lead to reductions in drivers' situational awareness and ability to negotiate complex traffic environments with the possibility that critical situations may arise. There is still work to be done to resolve the true extent, severity and incidence of cognitive distractions.

4.1.3 ADAS AND AUTONOMY

The increase in sophistication of IVIS has been accompanied by progress in the field of advanced driver assistance systems (ADAS). ADAS are designed to improve the safety and/or comfort of the driver and range from those that provide the driver with information about the current driving situation (such as Volvo's blind spot information system (BLIS)) to those that take active control of the vehicle (such as Mercedes-Benz' Active Lane Assist). As with IVIS, there are established guidelines for the evaluation and assessment of ADAS (e.g., RESPONSE 3 (Schulze et al., 2009)). However, the development of ADAS is reactive to dangers observed in driving—the support for the driver increases, as technology is developed to tackle threats to safety. The NHTSA (2012) draft guidelines on driver distraction state that the regulations it contains were deliberately not made mandatory to avoid unforeseen conflict with the introduction of new safety technologies that may permit the driver to engage in tasks previously considered excessively distracting. This acknowledges that tertiary tasks currently described as excessively distracting may become more acceptable if ADAS alleviates the task demand on the driver. Whilst the driver may be distracted, ADAS may provide additional hazard detection and appropriate responses such that the complete driver–vehicle system can still safely negotiate challenging driving situations.

An extreme implementation of ADAS is for the human driver to cede all control of the vehicle to technology, the concept of a fully autonomous vehicle. This is not a distant future concept but a present reality (e.g., Thrun et al. 2006; Levinson et al. 2011). However, whilst the technological challenges to fully autonomous cars are rapidly being surmounted, the legal issues surrounding accident liability in the event of a collision involving an autonomous vehicle are yet to be resolved (Beiker and Calo, 2010). The introduction of such a radical change in vehicle control must be achieved with minimal risk of danger or injury but this must be weighed against the potential beneficial effects that vehicle autonomy may bring. These include not only a reduction in injuries as a result of fewer driver error accidents but also sustainability benefits that may arise from more efficient use of vehicles. Whilst technological advances indicate that autonomous cars are close to market, the truth is that it will be several years before that scenario is realised. Furthermore, it is unlikely that the technology will be affordable to the majority of drivers for some time beyond its initial introduction such that widespread adoption in the vehicle fleet is many years away.

4.1.4 SUMMARY

A complex picture of driving risk emerges. The evidence indicates that:

> For the majority of the time, drivers believe they have the spare mental capacity to engage in tertiary, driving-unrelated tasks.

Technology is creating the opportunity for drivers to engage in a larger number of increasingly complex tertiary tasks.

The usability paradox indicates that well-designed, highly usable systems, although less distracting per interaction, may result in higher cumulative distraction if they are used more frequently.

When drivers are given the opportunity to engage in tertiary tasks, a proportion of drivers will take it.

Drivers' engagement in tertiary tasks causes changes in driving behaviour that results in an increase in the risk of collision.

The demands of the driving task are being modulated by the introduction of a variety of ADAS.

ADAS that reduce risk of driver error accidents to zero are unlikely to be widespread within this decade.

The way in which drivers manage competing demands from the driving task, from in-vehicle systems and by other life goals has parallels with the work of an influential military strategist from the 1960s.

4.2 BOYD AND THE OODA LOOP

In the Vietnam conflict, the capabilities of the aircraft of the United States Air Force (USAF) were called into question. Following the Korean War, the development of fighter aircraft had pursued goals of outright speed, range and armament at the expense of agility. This approach produced aircraft such as the North American F-100 Super Sabre, the McDonnell-Douglas F-4 Phantom and the Republic F-105 Thunderchief (Osinga, 2007). However, these were outclassed in Vietnam by the dedicated combat aircraft supplied by the Soviet Union.

Through the 1960s, US military strategist Col. John Boyd analysed aerial warfare tactics from the Korean War. He sought to discover why the USAF North American F-86 Sabre was successful in combat against the opposing Mikoyan-Gurevich MiG-15, which had superior speed and climb capabilities. Boyd suggested the principal advantages were that the cockpit configuration of the F-86 permitted better vision of the combat arena; F-86 pilots were better trained in the application of air-to-air tactics and so could respond more appropriately to the developing engagement; and that the F-86 had better control interfaces such that pilots could implement desired manoeuvres more readily. Based on these ideas, Boyd developed the concept of the Observation–Orientation–Decision–Action (OODA) loop (Thomas, 2010). Opponents in a dogfight must cycle through this loop, whether in an offensive or defensive position. Boyd suggested that a combatant who can get 'inside' the opponent's OODA loop, the combatant who can cycle through this loop more quickly and more effectively, is likely to be the victor (Boyd, 1986).

It has been asserted (Osinga, 2007) that the one reason for the poorer combat performance of USAF aircraft in Vietnam was because their designs (which predated Boyd's concept) did not enable rapid cycling through the OODA loop, leaving them vulnerable to more agile aircraft. Thereafter, Boyd's theory gained respect and had a

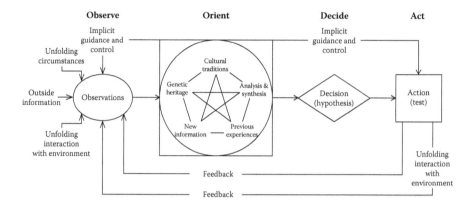

FIGURE 4.2 Boyd's OODA loop. (Adapted Richards, C. W. (2001), *A Swift, Elusive Sword: What if Sun Tzu and John Boyd Did a National Defense Review?* Center for Defense Information, May.)

significant influence on the design of combat aircraft. However, the OODA loop has applications beyond air warfare. Figure 4.2 shows the OODA loop.

The key statements that underpin OODA loop theory (adapted from Boyd, 1995) are as follows:

> Without our genetic heritage, cultural traditions, and previous experiences, we do not possess an implicit repertoire of psychophysical skills shaped by environments and changes that have been previously experienced.
>
> Without analyses and synthesis across a variety of domains or across a variety of competing/independent channels of information, we cannot evolve new repertoires to deal with unfamiliar phenomena or unforeseen change.
>
> Without a many-sided, implicit cross-referencing process of projection, empathy, correlation and rejection (across these many different domains or channels of information), we cannot even do analysis and synthesis.
>
> Without OODA loops embracing all of the above and without the ability to get inside other OODA loops (or other environments), we will find it impossible to comprehend, shape, adapt to and in turn be shaped by an unfolding evolving reality that is uncertain, ever-changing, and unpredictable.

Although derived with military operations in mind, these concepts can be applied usefully in the domain of driving. In its simplest format, controlling a moving vehicle requires the driver to use innate (e.g., Gibson and Walk, 1960) and learned (e.g., Rakison, 2005) abilities of motion perception to understand its movement through the environment. Perceptual information from primarily visual (e.g., Harris, Jenkin and Zikovitz, 2000) but also auditory (e.g., engine noise—Horswill and McKenna, 1999), proprioceptive (e.g., sense of self-movement and acceleration—Guedry, 1974) and possibly olfactory (engine/tyre smell) sensory channels must be *observed* by the driver. The movement of the driven vehicle within the environment and relative to other actors must be analysed and synthesised in the context of the goals of the driver (*orientation*). The driver must then *decide* how to respond based on the available information and on experience. The driver can engage a repertoire of appropriate learned *actions* to cause the vehicle to respond in the desired manner (see e.g., Groeger, 2000; Hole, 2007). It can be seen in Figure 4.2 that the decision step can be skipped by implicit processes.

The action of driving requires the driver to continually circulate around this loop to maintain safe control of the vehicle, responding to changes in the environment and in the status of the driven vehicle. However, a variety of factors can impair a driver's ability to cycle around the loop. Visual field impairments will affect a driver's ability to observe the unfolding circumstances of the driving situation with the potential to impair fitness to drive (Kotecha, Spratt and Viswanathan, 2008). A reduction in the ability to observe the situation will affect the driver's subsequent ability to orient, decide and act upon the available information. Szlyk et al. (1995) found that in simulator and on-road tests, drivers with age-related macular degeneration (AMD) showed a range of performance differences from a comparable control group but that these did not translate to an increase in real-world collision risk. The authors cited evidence of compensation by the AMD-affected drivers to reduce their exposure to risk by restricting driving to familiar areas, to slower speeds, to times of daylight and to simpler road configurations. This observation fits with the task difficulty homeostasis theory (Fuller, 2005) but can also be explained by less effective circulation of the OODA loop. Impaired observation reduces the quality of the output in all subsequent steps. Compared to a driver with normal vision, this would result in a driver experiencing greater difficulty achieving the same level of performance in a set driving task or the driver could experience the same level of difficulty by reducing the demands of the driving task—leading to the observed changes in driving exposure.

A driver under the influence of alcohol may suffer impairments to all stages of the loop. Alcohol has been shown to affect the ability to perceive visual stimuli (Moskowitz, 1973); to analyse and synthesise observed information (Peterson et al., 1990); to make timely and appropriate decisions (Mongrain and Standing, 1989; Burian, Liguori and Robinson, 2002) and to co-ordinate and implement motor skills (Kerr et al., 2004). The OODA loop forms a neat framework for placing the impairments to driving caused by alcohol in context.

4.3   OODA LOOPS AND DRIVER DISTRACTION

The US–EU Bilateral ITS Technical Task Force (2010) arrived at a definition of driver distraction as follows:

> Driver distraction is the diversion of attention from activities critical for safe driving to a competing activity.

Distractions fitting this description may impact on the driving task in different ways. Lee and Strayer (2004) presented a development of Michon's (1985) three-level temporal hierarchy over which driver distraction may occur. Operational behaviour refers to the second-by-second situation assessment and applied control inputs, an example of which is the driver's steering inputs to maintain position in the driven lane. Tactical behaviour operates over a time base of 5 to 60 seconds, an example of which might be moving into the appropriate lane on a highway in order to take the next exit. Strategic behaviour takes place over minutes to days and an example might be a decision by the driver to drive the vehicle in a fuel efficient manner—which will then have repercussions at the tactical and operational levels. Lee and Strayer's diagrammatic representation of this framework is shown Figure 4.3.

Each level of driving behaviour can be associated with an OODA loop. Activity at the higher levels cascades down to influence behaviour at the lower levels. Communication technologies enable a driver to remain engaged with OODA loops relating to life beyond the driving task. This may result in the driver's abilities of observation, orientation, decision and action to be impaired, leading to increased collision risk if the driver cannot compensate for this impairment (by e.g., reducing speed or stopping). The ubiquity of mobile telephony means that instant and unbroken communication is for many people the expected norm. The relatively low frequency of road accidents means that negative feedback on distracted driving (an

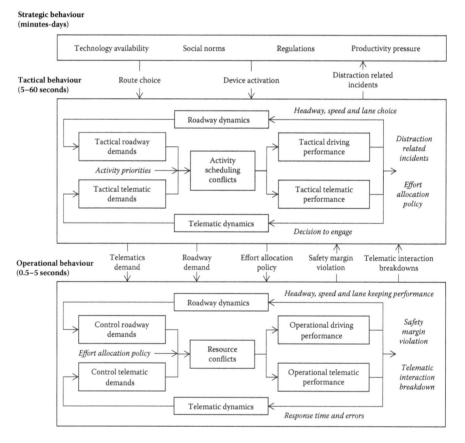

FIGURE 4.3 A framework to show how distraction can operate at three different levels of driving behaviour. (Adapted Lee, J. D. and Strayer D. L. (2004). Preface to the special section on driver distraction. *Human Factors*. Winter 46(4), 583–586.)

accident) is rare and drivers have an imperfect knowledge of how distraction impacts driving at an operational level. The OODA loops of driving may therefore take lower priority than the tertiary task, leading to the possibility of Engstrom's zombie behaviours and increased accident risk.

4.5   OODA AND ADAS

The driver is not the only element of the system where considering the OODA loop can be useful. Guidance about the design of crash warning interfaces for in-vehicle systems was summarised in the 2007 NHTSA report (Campbell, Richard, Brown and McCallum, 2007). This wide-ranging document synthesised documented evidence and expert judgment on guidelines for the design of such interfaces. It provides recommendations for the implementation of specific types of driving safety systems such as forward collision warning systems and lane change warning systems. However, the report begins by discussing general guidelines for the design of driver warnings. The warnings delivered by the safety systems are the culmination of:

The observation of a change in a particular metric.
The orientation of that observation by evaluating it against relevant thresholds.
The decision about what action to take based on that evaluation.
The action applied—to deliver a warning or not.

A human driver is subject to competing demands such that observation is not necessarily continuous and the mental resources for orientation and decision may be loaded by other tasks. An electronic safety system can continuously monitor the driving environment waiting to detect the specific conditions that the manufacturer has deemed constitute a safety risk and act as programmed should those conditions occur. If the system is sufficiently sensitive to the risk conditions, the OODA loop of the safety system will be 'inside' that of the driver. Assuming there are no negative behavioural adaptations to the presence of the system, this will result in improved safety of the car–driver system.

Aviation has benefitted from autonomous pilot support systems, such as autopilot for many years. Indeed, less than a decade after the Wright brother's first powered flight in 1903, the Sperry Corporation had created a system that used hydraulic actuators to maintain an aircraft on a fixed heading and in level flight. Such systems became commonplace after World War II and into the jet age. However, Parasuraman and Riley (1997) discussed the factors that limit the benefits that may be achieved by automation. A high incidence of false alarms may lead to disuse or under-utilisation of automation. The authors also describe automation abuse, in which the automation and implementation of functions occurs without due regard for the consequences for human performance. Beringer and Harris (2007) describe a number of reasons why autopilot systems may fail that are relevant to the driving situation. Insufficient training of the operator—pilots who have not been adequately trained in the use of autopilot systems can over-, under- or misuse the available technology. The lack of a conceptual model of the operation of the components of automation—pilots that do not understand the operation of the automated systems may come to rely on their intervention when it would be inappropriate to do so. Human performance limitations—the detection of malfunctions is influenced by limitations in visual and aural perception. Human factors and design issues—installed systems (and their associated warning signals) may not conform to standard human factors practices and principles. This evidence from the aviation world, where regimes for pilot fitness to fly, and training exceed those for driving, emphasises the need for thorough research and testing in the design of ADAS. Ill-conceived systems may fulfil one or more of these criteria, resulting in sub-optimal support to the driver. Understanding all the factors that underpin the OODA loop of the ADAS and how those might support or affect the OODA loops of the driver will help the automotive industry to develop safety systems that maximise driver benefit.

4.6 CONCLUSION

The technological development of the car in its early years was about improvements in ease of use and comfort. From the middle of the twentieth century, cars became more popular such that improving drivers' ability to handle vehicles in traffic situations, to avoid collisions and to mitigate outcomes when collisions occur became important aspects of their development. Later in the twentieth century, computer technology became sufficiently small, cheap and robust to be integrated into vehicles. This opened new horizons for information exchange and communication with both the driver and vehicle. Whilst the opportunities for the driver to remain contactable via mobile telephone and data links have expanded, the demands of the driving task have not grown simpler and the legal questions around the introduction of driver support systems that may alleviate the risk of distracted driving suggest that a complete technological solution to permit drivers to engage freely in driving unrelated tasks is some years away. Consequently, there is a dangerous transition period when the opportunities for drivers to engage in distracting tertiary activities are accelerating at a rate that is exceeding that of the development of ADAS systems that might help mitigate the effects of driver distraction.

Drivers may have many objectives that conflict with their discrete ability to control a vehicle. IVIS enable a driver to retain engagement with OODA loops that previously were distinct from driving situations. ADAS may compensate by operating within the OODA loop of the distracted driver. Considering the OODA loops that govern behaviour of the driver and of the driver–vehicle complex may enable an improved understanding of how distraction impacts performance, how driver support systems may mitigate distraction and improve the safety and efficiency of driving and to assist in the smooth implementation of autonomous vehicles.

REFERENCES

Beiker, S. and Calo, R. (2010). Legal Aspects of Autonomous Driving. (Workshop summary essay). Available at: http://me.stanford.edu/groups/design/automotive/ADP/Autonomous%20Driving%20-%20Essay%202010-12%20V0.5%28final%29.pdf. Accessed: February 2012.

Beringer, D. and Harris, H. (2007). *Automation in General Aviation: Two Studies of Pilot Responses to Autopilot Malfunctions.* (Technical Report No. DOT/FAA/AM-07/24). Washington, DC: Federal Aviation Administration, Office of Aerospace Medicine.

Boyd, J. R. (1986). tterns of Conflict (Unpublished briefing slides). Available at: www.dni-pogo.org/boyd/patterns.ppt. Accessed: January 2011.

Boyd (1995). The Essence of Winning and Losing; Summary briefing; Available at http://danford.net/boyd/essence.htm. Accessed: January 2012.

Breen, J. (2009). Car telephone use and road safety: Final report. An overview prepared for the European Commission. Available at: http://ec.europa.eu/transport/road_safety/specialist/knowledge/mobile/car_telephone_use_and_road_safety.pdf. Accessed: February 2012.

Burian, S. E., Liguori, A. and Robinson, J. H. (2002). Effects of alcohol on risk-taking during simulated driving. *Human Psychopharmacology*, 17, 141–150.

Campbell, J. L., Richard, C. M., Brown, J. L. and McCallum, M. (2007). *Crash Warning System Interfaces: Human Factors Insights and Lessons Learned* (No. HS 810 697). Washington, DC: National Highway Traffic Safety Administration.

Dingus, T. A., Klauer, S. G., Neale, V. L., Petersen, A., Lee, S. E., Sudweeks, J., Perez, M. A., Hankey, J., Ramsey, D., Gupta, S., Bucher, C., Doerzaph, Z. R., Jermeland, J. and Knipling, R. R. (2006). *The 100-Car Naturalistic Driving Study, Phase II—Results of the 100-Car Field Experiment.* DOT HS 810 593. Washington, DC: National Highway Traffic Safety Administration.

Engström, J. (2011). *Zombies i Trafiken: Effekter av Kognitiv Distraktion Påkörprestatation Och Olycksrisk.* Linköping, Sweden: Volvo Technology Transportforum.

Fuller, R. (2005). Towards a general theory of driver behaviour. *Accident Analysis and Prevention*, 37(3), 461–472.

Gibson, E. J. and Walk, R. D. (1960). The "visual cliff." *Scientific American*, 202, 67–71.

Groeger, J. A. (2000). *Understanding Driving: Applying Cognitive Psychology to a Complex Everyday Task.* Hove, UK: Psychology Press.

Guedry, F. E. (1974). Psychophysics of vestibular sensation. In *Handbook of Sensory Physiology, Vestibular System*, Kornhuber H. H. (ed.). Vol 1/2, pp. 3–154. Berlin: Springer-Verlag.

Harris L. R., Jenkin M. and Zikovitz D. C. (2000). Visual and non-visual cues in the perception of linear self motion. *Exp Brain Res* 135, 12–21.

Hole, G. (2007). *The Psychology of Driving.* Hove, UK: Erlbaum.

Horswill, M. S. and McKenna, F. P. (1999). The development, validation, and application of a video-based technique for measuring an everyday risk-taking behaviour: Drivers' speed choice. *J Appl Psychol* 84, 977–985.

Kerr, J. S., Sherwood, N., Hindmarch, I., Bhatti, J. Z., Starmer, G. A. and Mascord, D. J. (2004). The effects of alcohol on the cognitive function of males and females and on skills relating to car driving. *Human Psychopharmacology: Clinical and Experimental*, Vol. 7, (2), 105–114.

Klauer, S. G., Dingus, T. A., Neale, V. L., Sudweeks, J. D. and Ramsey, D. J. (2006). *The Impact of Driver Inattention on Near-Crash/Crash Risk: An Analysis Using the 100-Car Naturalistic Driving Study Data* (Report No. DOT HS 810 594). Washington, DC: National Highway Traffic Safety Administration, USDOT. Available at: http://www.nhtsa.gov/DOT/NHTSA/NRD/Multimedia/PDFs/CrashAvoidance/2006/DriverInattention.pdf. Accessed: February 2012.

Klemens, G. (2006). *The Cellphone: The History and Technology of the Gadget That Changed the World.* London: McFarland.

Koch, C. (2004). *The Quest for Consciousness: A Neurobiological Approach.* Englewood, CO: Roberts and Company Publishers.

Kotecha, A., Spratt, A. and Viswanathan, A. (2008). Visual function and fitness to drive. *Br Med Bull.* 87, 163–74.

Lee, J. D. and Strayer D. L. (2004). Preface to the special section on driver distraction. *Human Factors*. Winter 46(4), 583–586.

Levinson, J., Askeland, J., Becker, J., Dolson, J., Held, D., Kammel, S., Kolter, J., Langer, D., Pink, O., Pratt, V., Sokolsky, M., Stanek, G., Stavens, D., Teichman, A., Werling, M. and Thrun, S. (2011). Towards fully autonomous driving: Systems and algorithms. In *Proceedings of the IEEE Intelligent Vehicles Symposium* 2011 (IV11). Baden-Baden, Germany.

McKnight, J., and Adams, B. (1970). *Driver Education and Task Analysis, Vol. 1* (Technical Report). Washington, DC: National Highway Safety Bureau.

Michon, J. A. (1985). A critical view of driver behavior models: What do we know, what should we do? In L. Evans and R. C. Schwing (eds.), *Human Behavior and Traffic Safety*, pp. 485–520. New York: Plenum.

Mongrain, S. and Standing, L. (1989). Impairment of cognition, risk-taking, and self-perception by alcohol. *Percept Mot Skills*. August 69 (1):199–210.

Moskowitz, H. (1973). Laboratory studies of the effects of alcohol on some variables related to driving. *Journal of Safety Research*, Vol. 5(3), September, 185–199.

National Highway Traffic Safety Administration (2012). Visual-Manual NHTSA Driver Distraction Guidelines for In-Vehicle Electronic Devices. NHTSA-2010-0053. Available at: http://www.nhtsa.gov/staticfiles/rulemaking/pdf/Distraction_NPFG-02162012.pdf. Accessed: February 2012.

Osinga, F. (2007). *Science, Strategy and War: The Strategic Theory of John Boyd*. Abingdon, UK: Routledge.

Parasuraman, R. and Riley, V. (1997). Humans and automation: Use, misuse, disuse, abuse. *Human Factors*, 39, 230–253.

Peppers, D. and Rogers, M. (2008). *Rules to Break and Laws to Follow: How Your Business Can Beat the Crisis of Short-Termism*. New York: Wiley.

Peterson, J. B., Rothfleisch, J., Zelazo, P. D., and Pihl, R. (1990). Acute alcohol intoxication and cognitive functioning. *Journal of Studies on Alcohol*, 51, 114–122.

Rakison, D. H. (2005). Developing knowledge of motion properties in infancy. *Cognition*, 96, 183–214.

Richards, C. W. (2001). *A Swift, Elusive Sword: What if Sun Tzu and John Boyd Did a National Defense Review?* Center for Defense Information, Washington DC, USA.

Schulze, M., Mäkinen, T., Irion, J., Flament, M. and Kessel, T. (2009). RESPONSE 3: Code of Practice for the Design and Evaluation of ADAS, Version 5, Preventive and Active Safety Applications Integrated Project, EU IST contract number FP6-507075. Available at http://www.ocea.be/images/uploads/files/20090831_Code_of_Practice_ADAS.pdf. Accessed February 2012.

Stevens A. (2008). European approaches to principles, codes, guidelines and checklists for in-vehicle HMI. In: *Driver Distraction Theory Effects and Mitigation*, (eds.), M. A. Regan, J. D. Lee and K. L. Young. Boca Raton, FL: CRC Press.

Stevens, A., Quimby, A., Board, A., Kersloot, T. and Burns, P. (2002). Design Guidelines For safety of In-Vehicle Information Systems. Project report PA3721/01. Crowthorne, United Kingdom: Transport Research Laboratory.

Strayer, D. L., Watson, J. M. and Drews, F. A. (2011). Cognitive distraction while multitasking in the automobile. In B. Ross, (ed.), *The Psychology of Learning and Motivation*, Vol. 54, Burlington: Academic Press, pp. 29–58.

Stutts, J., Feaganes, J., Reinfurt, D., Rodgman, E., Hamlett, C., Gish, K. Staplin, L. (2005). Driver's exposure to distractions in their natural driving environment, *Accident Analysis and Prevention*, Vol. 37, (6) November, 1093–1101.

Szlyk, J. P., Pizzimenti, C. E., Fishman, G. A., Kelsch, R., Wetzel, L. C., Kagan, S. and Ho K. (1995). A comparison of driving in older subjects with and without age-related macular degeneration. *Arch Ophthalmol*. August, 113(8),1033–40.

Thomas, Lt-Col. J. (2010). Abandoning the Temple—John Boyd and Contemporary Strategy. *Australian Army Journal*, Vol.VII (3). Available at: http://www.army.gov.au/Our-future/DARA/Our-publications/Australian-Army-Journal/~/media/Files/Ourfuture/DARApublications/AAJ_Summer_10.ashx. Accessed: January 2012.

US–EU Bilateral ITS Technical Task Force (2010). Expert Focus Group on Driver Distraction: Definition and Research Needs. Available at: http://ec.europa.eu/information_society/activities/esafety/doc/intl_coop/us/eg_driver_distract.pdf. Accessed: January 2012.

Thrun, S., Montemerlo, M., Dahlkamp, H., Stavens, D., Aron, A., Diebel, J., Fong, P., Gale, J., Halpenny, M., Hoffmann, G., Lau, K., Oakley, C., Palatucci, M., Pratt, V., Stang, P., Strohband, S., Dupont, C., Jendrossek, L.E., Koelen, C., Markey, C., Rummel, C., Van Niekerk, J., Jensen, E., Alessandrini, P., Bradski, G., Davies, B., Ettinger, S., Kaehler, A., Nefian, A. and Mahoney, P. (2006). Stanley: The robot that won the DARPA grand challenge, *Journal of Field Robotics*, Vol. 23, September, 661–692.

World Bank (2012). Mobile Cellular Subscriptions (Per 100 People). Washington, DC. Available at: http://data.worldbank.org/indicator/IT.CEL.SETS.P2. Accessed: February 2012.

5 Ergonomics Issues with Advanced Driver Assistance Systems (ADAS)

Mark S. Young
Brunel University, UK

CONTENTS

5.1   ADVANCED DRIVER ASSISTANCE SYSTEMS (ADAS)

We have come a long way in the development of vehicle technology since the humble horseless carriage of the late 19th century. The first decade of the 21st century has seen particular advancement in active safety systems—those which are designed to prevent, or mitigate the consequences of collisions by taking automatic control of the vehicle in some way (as opposed to passive safety, such as seatbelts and airbags, which prevent or mitigate the consequences of injuries resulting from a collision). Such advanced driver assistance systems (ADAS) purport to improve aspects of safety, comfort and convenience (e.g., Richardson et al., 1997) by either supporting the driver or taking over certain driving tasks. As such, ADAS devices can be contrasted with in-vehicle information systems (IVIS), which—as the name suggests—provide information to the driver and might not necessarily be related to the driving task. Whilst the terminology is by no means universal, the current chapter focuses on ADAS devices as outlined here; IVIS devices are the subject of another chapter in this book.

The kinds of ADAS devices we are interested in here cover those that provide warnings or advice, as well as those that intervene or control the vehicle in some way. In many ways, this progression reflects the evolution of the technology—being implemented in the first instance as a warning system, until confidence grows in the reliability of the technology and it becomes increasingly used to intervene or assume control of the vehicle. Moreover, the systems on offer both now and in the near future deal more or less explicitly with aspects of safety. The key point, though, is that all of these systems are aimed at influencing vehicle control—either directly or via the driver.

Many of these devices will be familiar to today's drivers, while others are still in the pipeline. Indeed, automatic gearboxes could be seen as an example of ADAS, and these have been around since the 1940s. Similarly, conventional cruise control was invented in the 1950s. These are, in many ways, the original 'comfort and convenience' devices for drivers. The evolution of cruise control into adaptive cruise control (ACC) did not arrive until the end of the 20th century. ACC goes beyond conventional cruise control by not only maintaining your car's speed at a set value, but also adjusts speed to maintain distance from vehicles in front by using a microwave radar to detect other vehicles in your lane. Whilst early versions of ACC had limited braking authority, being designed to work only at cruising speeds, in 2007 a 'stop-and-go' capability was introduced to ACC, which could bring your car to a standstill (thus enabling the use of ACC in traffic queues).

Nevertheless, ACC remains very much a 'comfort and convenience' system. Meanwhile, other devices throughout history have been more explicitly aimed at improving safety. The early 1970s saw the introduction of antilock braking systems (ABS), which use rapid cadence braking to prevent wheels locking up under extreme braking. By a similar token, electronic stability control (ESC), which has been around since the mid-1990s, detects skids in cornering manoeuvres and applies braking (or power) individually to the four wheels in order to correct the skid and maintain control.

Whilst ABS and ESC both apply a level of rapid corrective inputs that would be impossible for a human driver to achieve, more recent safety systems influence vehicle control at a much more conscious level. Taking ACC a step further, forward collision warning systems were introduced in 2006. These use similar radar systems to detect an impending collision with a vehicle in front, and alert the driver to it. Moving from warnings to actual vehicle control, the following year saw the extension of forward collision warnings to include automatic braking, and several manufacturers now offer collision mitigation braking systems (CMBS). Although at the time of writing, these have limited braking authority (as with ACC); there are systems already available which guarantee that they will prevent a collision at low speeds. Such systems will still brake at higher speeds, which will mitigate the consequences of a crash, but they cannot guarantee it will be avoided.

Turning to lateral control, the first lane departure warning systems were introduced at the turn of the 21st century. These use on-board cameras and image processing technology to detect lane markings, and provide warnings to the driver if the vehicle is crossing one of these lines. The warnings vary between manufacturers, but are usually either auditory or haptic (e.g., vibrating the steering wheel or seat to provide a 'virtual rumble strip'). Typically, the warning will be cancelled if the driver is using the turn signals to indicate an intended lane change. Similarly, blind spot warning systems, introduced in 2005, use cameras to detect vehicles in the driver's blind spot when a turn has been indicated, and alert the driver usually with a visual warning in the relevant side mirror. The logical next step from lane departure warnings is to use the data gathered from the camera systems instead to control the vehicle, and such lane keeping systems have been available since around 2006. Many of these do not actually assume full steering control of the vehicle, instead providing some haptic feedback on the steering wheel (e.g., increased resistance) to gently nudge the driver back into lane.

In the near future, we can expect more advanced collision warning systems (with enhanced braking authority) and intelligent speed advisory (ISA) systems. The latter systems use Global Positioning System (GPS) sensors to compare a driver's speed and location with a stored map database of speed limits. If the vehicle is exceeding the speed limit, then the driver can either be warned via an on-board interface (visual, auditory or haptic—such as increased resistance on the accelerator pedal), or in more extreme implementations, the engine control unit can cut power to regulate speed. Although barriers exist to the implementation of ISA at present (namely the lack of detailed and up-to-date map databases for speed limits), manufacturers are already offering advice on speed limits using speed sign recognition cameras to present a constant reminder of the speed limit on the instrument panel. Advances in camera and display technology are also allowing drivers to be presented with all-round vision, and even night-vision displays for low visibility. Some of these systems can even detect and recognise hazards and obstacles (such as pedestrians) to further refine the aforementioned collision warning and avoidance systems.

As well as systems to monitor the surrounding environment, devices which monitor the driver are also being developed. Fatigue is a particular issue, and some systems are already on the market that monitor and analyse driver behaviour to detect patterns of steering control that have been associated with tired drivers. When the system detects a threshold level of these behaviours, the driver is given a warning in the instrument cluster. Other systems will soon be available that use eye-tracking cameras embedded in the dashboard, in order to detect signs of fatigue. Such technology can also be used to monitor driver distraction, providing auditory or haptic warnings if the driver's gaze is diverted from the primary attention zone for too long. In-vehicle distraction can be managed as well, with systems that can postpone or suppress low-priority messages or telephone calls if driver workload is deemed to be too high (e.g., Broström, 2006; Buchholz, 2003). Workload might be derived from driver behaviour (steering, acceleration and braking inputs), eye or head tracking, or environmental cues (e.g., GPS data).

In further developments of such adaptive systems, Smith et al. (2009) describe an innovative collision warning system that monitors the driver's head position and adapts its warnings depending on whether the driver is watching the road or not. The system could adapt its warnings either positively (i.e., trying to attract the attention of a distracted driver by either presenting the warning earlier, in a different location, or through a different modality) or negatively (i.e., attenuating a warning so as not to annoy an attentive driver), to increase safety or acceptance, respectively.

Other motivations for ADAS development include social and environmental concerns. As well as developments in low carbon vehicle technologies, more recently the market has seen a number of 'green' ADAS interfaces aimed at encouraging environmentally friendly driving—or 'eco-driving'. Various types of existing ADAS devices can have ancillary benefits for eco-driving, while specific products for eco-driving assistance are beginning to emerge (e.g., Ericsson, Larsson and Brundell-Freij, 2006; Van der Voort, Dougherty and Van Maarseveen, 2001; Van Driel, Hoedemaeker and Van Arem, 2007). For instance, some satellite navigation systems now offer an 'economical route' planning option, alongside 'fastest route' or 'shortest route'. Young and Birrell (2012) described the development of 'Foot-LITE', a driver monitoring system which provides feedback on driving style to encourage both safe and eco-driving.

Meanwhile, driver assistance could be used to support the mobility needs of groups of drivers with reduced capabilities, such as older drivers (e.g., Young and Bunce, 2011). Many of the technologies discussed above could support the cognitive functioning of older drivers. Nevertheless, such systems are very much a result of technology 'push' rather than user 'pull'; what is needed is a balanced, user-centred assessment of these technologies. A recent UK project explored this very issue, and reported that most new in-car technologies have so far ignored older drivers' needs (Haddad and Musselwhite, 2007). Research suggests (e.g., Bradley, Keith and Wicks, 2007; Keith et al., 2007) that technological assistance inside the car will only be of benefit if it has been designed from a user-centred perspective. In particular, the diminished capacities of older drivers could render them more susceptible to overload with poorly designed assistance (cf. Harvey et al., 1995; Lundberg, 2003).

Across all of these ADAS products, there is mounting evidence that technological driver assistance can improve safety and reduce road fatalities. However, such benefits will only ever be fully realised if the system design takes into account the human factor behind the wheel, especially as these devices impinge more and more on the driver's task.

Where traditional automatic systems (such as automatic transmission and conventional cruise control) have sought to assume the lower-level, operational components of vehicle control, new technologies are taking over more tactical and even strategic aspects of driving (cf. Ranney, 1994). Young, Stanton and Harris (2007) explicitly distinguished between these levels as 'vehicle automation' and 'driving automation' respectively. For instance, conventional cruise control (CC) simply adjusts the throttle to maintain speed. On the other hand, adaptive cruise control (ACC) removes a cognitive task from the driver—perceiving speed of a lead vehicle, deciding whether to adjust speed in response, and taking appropriate action. Collision avoidance (CAS) and collision warning systems (CWS) take this a step further, by making a potentially stressful decision about whether to take emergency action. Even lane-keeping systems or Active Steering (AS), which might appear to be an example of vehicle automation, relieve the driver of a significant cognitive workload (Young and Stanton, 2002), owing to the fact that steering is a second-order tracking task (Wickens, Gordon and Liu, 1998).

It seems, then, that more and more driving tasks (as opposed to vehicle control tasks) are falling within the capabilities of automation. Whilst at present the technologies are working independently of each other, full integration of these systems will make an autonomous vehicle a commercial reality. Futurologists and ergonomics researchers alike are predicting that by 2030 fully automated vehicles will be on our roads (e.g., Walker, Stanton and Young 2001). Whilst the engineering of these vehicles seems on track, the understanding of the interaction between the computing, the vehicles behaviour and the drivers' reactions seem much less clear. In other words, automotive automation has wide implications for driver human factors, with the potential to affect situation awareness, mental workload, and driver stress (Stanton and Young, 2000, 2005).

5.2 ERGONOMICS IMPLICATIONS

Whether the ADAS takes the form of warning the driver, or intervening in vehicle control, there are significant ergonomics concerns regarding the impact on driver behaviour and performance. A warning device, by virtue of its adding information to the driving task, has implications for driver mental workload and distraction—and distraction is a causal factor in accidents (e.g., Regan, Lee and Young, 2009). Conversely, an automated control system could excessively reduce workload, or affect the driver's situation awareness. Moreover, drivers could become overdependent on the system and complacency could creep in, as they believe the vehicle has become inherently safer with the new technology.

Much of the ideology behind driving automation has been around for some time in aviation systems, and such lessons can be—and, indeed, have been—transferred from the aviation domain (Billings, 1993; Stanton and Marsden, 1996; Wiener and Curry, 1980). It has been noted (Stanton and Marsden, 1996) that automation has been implicated in a number of fatal aviation accidents. Root cause analysis by accident investigators has identified psychological factors such as boredom and inattention under conditions of low workload, cognitive strain under conditions of very high workload, failure of automated systems to meet pilots' expectations, and overreliance on the technology. There is no reason to believe that these factors are unique to the domain of aviation automation and, if poorly designed, they could transfer to driving automation. The challenge for ergonomics is to ensure that the speed of technological development does not outpace that of the human brain.

5.2.1 MENTAL WORKLOAD

Mental workload is a particular issue with ADAS devices. Driver overload with an additional task or interface in the vehicle can adversely affect performance (Donmez, Boyle and Lee, 2007; Horberry et al., 2006), particularly if workload is already high (e.g., in urban driving; Liu and Lee, 2006; or in abnormal or emergency scenarios) or if the driver has a lower capacity to respond (e.g., in the elderly; May, Ross and Osman, 2005; or if the driver has less skill or experience). Studies have shown that while conducting a difficult cognitive task (such as maths addition), drivers spend less time looking at areas in the peripheries (such as mirrors and instruments) and instead focus on looking centrally ahead (Harbluk et al., 2007). Even though time looking outside of the vehicle remained unchanged, these results suggested a change in drivers' allocation of attention.

Whilst the presence of such a secondary task can increase the potential risk of an accident or incident, it has been suggested that drivers may have up to 50% spare visual capacity (Hughes and Cole, 1986) during 'normal' driving, suggesting that some secondary tasks may be able to be conducted with no subsequent increase in crash risk. Therefore, it is thought that other contributing factors also have to occur at the same time for the risk to manifest itself (Angell et al., 2006). Contributing factors may include the presence of a junction, urban driving or unexpected events. Such factors can impair the reactions of a distracted or overloaded driver since their spare attentional capacity has been absorbed by the secondary task. With the increasing prevalence and potential of new IVIS products coming to market, this spare capacity could soon get consumed, thus creating distraction issues if not carefully managed.

On the other hand, there is some evidence that ADAS in the form of vehicle auto-mation, such as adaptive cruise control (ACC), can reduce driver mental workload in certain situations. Bar-Gera and Shinar (2005) suggested that car following and headway monitoring is a demanding task, and that devices such as ACC can relieve these demands. Likewise, Ma and Kaber (2005) argued that ACC relieves mental workload and hence improves situation awareness, which in turn enhances perfor-mance. They used a medium-fidelity driving simulator to show that ACC reduced workload and improved driving performance in terms of speed, headway and lateral variability. Similarly, Stanton, Young and McCaulder (1997) found the ACC system caused a significant reduction in mental workload on a secondary task measure. However, other studies report no effects of ACC on mental workload (e.g., Nilsson, 1995; Ward, Fairclough and Humphreys, 1995; Young and Stanton, 2002).

But reductions in workload can go too far, and mental underload can be just as detrimental to performance as mental overload. Young and Stanton (2002) argued that performance degrades in underload situations due to a shrinkage in attentional capacity, under their Malleable Attentional Resources Theory (MART). If mental workload is lowered enough, then the consequent resource shrinkage might prove too much to allow the operator to cope with a sudden increase in demand (such as automation failure). The general consensus is that mental workload optimisation is crucial to maintaining effective task performance.

Young and Stanton (2002) investigated ACC with the use of an active steering (AS) system in a driving simulator study. Although ACC had little effect on mental workload, AS did significantly reduce workload—and this was also reflected in a reduction in attentional capacity, as predicted by their MART theory. The study was later extended to include drivers of different skill groups (Young and Stanton, 2007b), and interestingly it was found that less skilled drivers did show a reduction in mental workload with ACC. Furthermore, the presence of AS seemed to improve the longitudinal performance of less skilled drivers. Thus, the skills gap was attenuated as more levels of automation were introduced (cf. Shinar, Meir and Ben-Shoham, 1998; Ward, 2000). Contrary to expectations, then, reductions in workload were associated with improvements in performance for unskilled drivers—despite the fact that all skill groups were susceptible to resource shrinkage in underload conditions.

Rather than facing a possible adverse situation of underload with automation, then, drivers with less skill are evidently being overloaded under normal (manual) conditions, and could thus ostensibly benefit from the introduction of automation.

Nevertheless, driving performance for skilled drivers was largely unaffected by automation. However, consider the nature of the task used in these studies—a highly controlled, normal driving scenario. That is, there were no emergency or abnormal events. So, Young and Stanton (2001) tested the same set-up, but included a failure of the ACC system towards the end of the trial. Now a detrimental effect of underload was observed, especially for unskilled drivers, and especially when they knew the automation might fail—only half of these participants responded to the failure event, compared to most in the normal workload condition. Other research into performance with automation has only found detrimental effects when there is a sudden increase in demand, such as in an emergency scenario. For instance, a test-track study by Rudin-Brown and Parker (2004) found that whilst ACC reduced workload, this was associated with increased reaction times to a hazard detection task and fewer safe braking interventions by drivers.

5.2.2 SITUATION AWARENESS

A lack of situation awareness has been associated with passive supervision of automated systems, or what is known as the 'out-of-the-loop' performance problem (e.g., Endsley, 1995). Being 'out of the loop' degrades a driver's perception, understanding and prediction of the situation as it unfolds, again with impacts on performance. This out-of-the-loop performance problem can be manifest in vigilance failures (Molloy and Parasuraman, 1996), or difficulties in recovering control following automation failure (Endsley and Kiris, 1995).

Operators need to be aware not just of where they are and what is happening, but also of what the system is doing. One of the most prevalent problems with automation, particularly in aviation, is mode errors (Stanton and Marsden, 1996). The simplest example of a mode error is attempting to set the time on a digital clock, when the clock is actually in alarm mode. A moded system can offer increased functionality and flexibility, but complex, event-driven systems may change modes without input from, or feedback to, the operator. This can cause confusion and increased cognitive demand as the user tries to keep track of mode transitions and the system state (Sarter and Woods, 1995). Consequently, 'automation surprises' may occur, in which the system behaves according to specifications, yet this is quite different to that which the operator expects or desires. In these circumstances, autopilot failures can lead to overcompensation and wild oscillatory responses, which may lead to loss of control and a consequent crash.

Automation surprises are also often determined to be the cause of aviation accidents involving modern 'glass cockpit' aircraft. The good safety record of these aircraft often leads the authorities to attribute errors to pilot training or procedures. However, an analysis of automation incidents found that such errors are more frequent and severe in glass cockpit aircraft (Kantowitz and Campbell, 1996). Many incidents are due to a lack of feedback from the system, to the extent that even pilots experienced with automation are sometimes surprised. Feedback was also implicated when it was found that mode confusions are often only detected by observing the system response, rather than the automation displays (Palmer, 1995). Some suggest that instead of fighting the computer, pilots should occasionally switch it off and look out of the window. Moreover, there is a long-standing consensus that automation can lead to skill degradation over time, such that operators do not know how to reclaim control when necessary (e.g., Bainbridge, 1982; Parasuraman, 2000).

5.2.3 TRUST

Excessive trust in automation has been associated with vigilance failures (Molloy and Parasuraman, 1996; Parasuraman et al., 1996), and trust in automation is an issue which has attracted special attention (e.g., Muir, 1994; Muir and Moray, 1996). Depending on the reliability of the automated system, trust could either be too high (leading to complacency) or too low (resulting in the system being switched off and negating its benefits). As with workload, maintaining trust is a delicate balance.

Trust is governed by self-confidence, confidence in the system, and the reliability of the system (Hancock and Parasuraman, 1992), and the level of trust in an automated system determines the human's use and monitoring of that system. Lee and Moray (1994) found that automation tends to be used when trust in the device exceeds operators' self-confidence in their own performance at the task. If self-confidence outweighs trust, manual control will prevail. However, appropriate use may be upset, if operators excessively rely on the machine (leading to misuse), if false alarms reach an excessive rate (leading to disuse), or if the operator is neglected by the system (resulting from abuse of automation in design; Parasuraman and Riley, 1997).

The key issues with ADAS devices, particularly collision warning type systems, are driver trust and false alarms (Lees and Lee, 2007), and behavioural adaptation (e.g., Cacciabue and Saad, 2008). If the driver comes to rely on the system as a 'safety net', it may influence their driving towards more risky behaviours, and thus compromise safety more than without the system. Indeed, a classic case of behavioural adaptation is in drivers treating an adaptive cruise control (ACC) system as if it were a collision avoidance device (cf. Rudin-Brown and Parker, 2004)—essentially abusing the system beyond its design envelope. But as with all such human–machine interactions, whilst it is tempting to blame the human for such behaviour, it is actually the design of the system that has 'encouraged' the use of it in this way.

Rudin-Brown and Parker (2004) found that drivers failed to detect a failure of ACC for an average of 23 seconds, and concluded that drivers' behavioural adaptation to ACC would reduce its effectiveness in preventing rear-end collisions by 33%. Similarly, Stanton, Young and McCaulder (1997) used a driving simulator to explore the effects of ACC failure on driver performance. Participants were required to follow a lead vehicle with ACC engaged. At a predetermined point, the ACC system would fail to detect the lead vehicle braking, necessitating participant intervention to avoid a collision. It was found that one-third of all participants collided with the lead vehicle when ACC failed. However, there was no control condition involving a critical situation without the use of ACC. Nilsson (1995) also investigated the effects of

ACC in critical situations. It was found that ACC did influence behaviour, such that for the situation in which collisions occurred (when the car approached a stationary queue), 80% of the collisions occurred when ACC was engaged. Nilsson attributed this to the expectations that drivers have about ACC, rather than to changes in workload or alertness.

Young and Stanton (2007a) conducted a driving simulator experiment assessing brake reaction times of skilled and unskilled drivers under two different levels of automation. When compared to previous data gathered during manual driving, there seemed to be a striking increase in reaction times for these automated conditions. Increased brake reaction times when using ACC in critical situations have been observed elsewhere (Hogema et al., 1997). Implications for the design and safety of automated vehicle systems were discussed by Young and Stanton (2007a). Since ACC and other longitudinal control devices are primarily aimed at reducing headway in order to increase road capacity, it seems ironic that the evidence suggests drivers actually need more time to react in emergency situations. ACC designers face a dilemma of defining safe headway in terms of the vehicle's capabilities or the driver's reaction times (cf. Taieb-Maimon and Shinar, 2001).

A study of an emergency situation in an automated highway system found that only half of the drivers reclaimed control (de Waard et al., 1999). The remainder were faced with a distance headway as low as 10 centimetres. The authors claimed that this was an optimistic estimate, with the simulated environment essentially getting the best performance out of their participants.

5.3   AUTOMATION PHILOSOPHIES

Whilst it is as well to be aware of these ergonomics issues with ADAS, the challenge for designers is to ensure that such systems are designed to avoid these problems while maximising the benefits. Technological interventions can take any form from full operational control, to simple decision support or task assistance (Kaber and Endsley, 1997; Labiale, 1997). Determining the right level of automation for a task can help to optimise driver workload, situation awareness, performance and satisfaction. However, we can also take a step back from these issues and consider a similar question at more of a macro level. That is, what is our guiding philosophy when implementing automation?

In aviation, there have emerged two different philosophies regarding the authority of automation: hard and soft automation. Hard protection exists to prevent the pilot from inadvertently exceeding safety limits. The rationale behind hard protection is largely to protect the airframe—if the pilot should inadvertently take the aircraft beyond its performance envelope, automatic interventions will prevent damage and maintain flight dynamics. Hard automation, then, employs the technology to prevent error; as such, it has ultimate authority and can override the human operator's inputs.

Soft protection, on the other hand, uses automation as a tool to aid pilots, giving them full authority to override the automated systems if they want (or need) to. The pilots therefore have access to the full performance envelope and will not be overridden by the automated systems. There are still automated advisories in this soft protection scheme—if the pilot wishes to exceed set limits, s/he is required to apply more force than normal on the controls (Hughes and Dornheim, 1995).

Hard and soft automation therefore use similar sensors and control devices, but to different ends. Hard automation takes the pilot's input, determines whether it is sensible, and if necessary takes its own action before passing the instructions on to the control surfaces. This can be beneficial in certain situations. A good example is if the pilot has received a collision warning and, in a panic reaction, pulls hard back to gain altitude. Without an associated increase in thrust, the aircraft would soon stall. In that situation, the aircraft will itself apply the necessary amount of thrust to climb without stalling. There are circumstances, though, in which this level of computer authority can cause problems rather than resolving them. The crash of an A320 at an air show near Paris in 1988 was caused because the automation had made an incorrect assessment of the pilot's inputs (Beaty, 1995). Making a low-level fly-by with the undercarriage down, the computer assumed the pilot wanted to land and so throttled back the engines. When the pilot attempted to pull clear, the necessary thrust was not available, and the aircraft plunged into woodland at the end of the runway.

Soft automation makes a similar assessment of pilot inputs, but will only give feedback if the control requests appear to represent a safety risk. If the pilot persists, the soft automation will then pass the inputs directly to the control surfaces without intervention. Again, there are certain situations in which the pilot may legitimately wish to take the airframe beyond its performance limits. An incident involving an engine failure on a Boeing 747 in 1985 was only recovered after the aircraft had lost 30,000 feet in an uncontrollable dive (see Norman, 1990). Needless to say, the airframe was significantly stressed during both the descent and the recovery, and substantial damage was caused. Interestingly, though, if that aircraft had been fitted with a hard protection system, the pilot would not have been able to recover control. Both philosophies have advantages, then.

In anticipating the degree of vehicle automation that might become standard issue in the future, it is wise to consider the question of whether hard or soft automation provides the best solution for road vehicles. As with the examples above, hard automation will overrule the driver if s/he exceeds the vehicle protection envelope, whereas soft automation will allow the driver to override it, and have access to the full operating limits of the vehicle.

Before looking forward, though, let us review existing systems from this perspective. The automatic gearboxes already discussed are primarily categorised as hard automation—whilst the driver may usually make limited gear selections (e.g., the use of 'kickdown' or rudimentary gear lever settings), in the main, the choice of gear is decided by the automation. Anti-lock braking systems (ABS) are similar—leaving aside the possibility to arm or disarm the system, an ABS intervention is made purely on an assessment of the vehicle's braking dynamics. Conventional cruise control, on the other hand, can be classified as soft automation—the driver decides how and when to set the system, and can resume control at any time.

Moving on to the more advanced technologies available at present and in the near future, ACC and AS represent examples of soft protection, in that they are selectable by the driver and any manual control inputs will override them. Similarly, collision warning systems offer information and advice to the driver without necessarily

TABLE 5.1

Matrix of Hard and Soft Automation Categories against Vehicle and Driving Automation Types

	Hard Automation	Soft Automation
Vehicle automation	Auto transmission	CC
	ABS	
	Traction control	
	ESP	
Driving automation	CAS	ACC
	ISA	AS
		CWS
		Parking aids

Source: Adapted from Young, M. S., Stanton, N. A. and Harris, D., (2007), Driving Automation: Learning from Aviation about Design Philosophies. *International Journal of Vehicle Design*, 45(3), 323–338.

assuming control—similar to the soft protection systems in aircraft. Conversely, a collision avoidance system, set up to intervene automatically in an impending collision, is more akin to hard protection. Similarly, the intelligent speed adaptation (ISA) systems discussed earlier in this chapter can fall into either camp. A system which warns the driver that they are exceeding the speed limit, perhaps giving extra resistance on the accelerator pedal, would represent soft automation. The system, which overrides the driver's input and will not allow them to break the speed limit, though, is a hard automation system—and some authors have expressed concern about this implementation (e.g., Young, Stanton and Harris, 2007).

The orthogonal classification of automation systems into either vehicle or driving automation, noted earlier in this chapter, allows us to observe trends in ADAS development (see Table 5.1). Both vehicle and driving automation systems can be designed for soft or hard protection, but generally the ergonomics concerns are focused on the higher-level driving automation systems, rather than vehicle automation. So, we may seek to classify these cognitive problems according to whether the system falls into the hard or soft automation category.

Issues of mental workload have been identified with some soft automation systems. In one study (Young and Stanton, 2002), AS significantly reduced driver mental workload, and the consequent underload may lead to performance problems if and when the driver needs to reclaim control. Similarly, it has been suggested (Landau, 2002) that a proliferation of driver support systems could overload the driver, thus nullifying any stress or satisfaction benefits of each individual system.

Hard driving automation, on the other hand, is largely associated with problems of trust and situation awareness. If the system is designed to assume control with little input from or feedback to the driver, then the driver may have difficulty in developing situation awareness of its operation in a given scenario. Without knowing exactly how it might behave, the driver could become distrustful of the system (i.e., lack of trust) or even develop misplaced trust (i.e., over-trust or complacency; Parasuraman and Riley, 1997). Then, the driver's situation awareness will be inadequate or inappropriate, resulting in potential performance problems in a critical situation.

Although somewhat coarse, this analysis indicates that the range of cognitive and performance problems seems to be more severe when implementing driving automation in the hard protection category. However, the cognitive factors involved when interacting with automation are all interdependent (Stanton and Young, 2000), and so we also need to consider the level of interaction between driver and automation.

There is widespread agreement in ergonomics that, until automation is good enough to replace the human completely, the philosophy behind automation should be one of supporting, not replacing the operator, in the same way as a human co-pilot or co-driver would (e.g., Hoc, Young and Blosseville, 2009; Young, Stanton, and Harris, 2007). Only then will these technologies realise their objectives of improving driver satisfaction and safety. In response, research in this area has been directed towards optimising the cooperation of the automation with the operator at various levels of the task (e.g., Hoc, 2001; Hoc and Blosseville, 2003; Hoc and Lemoine, 1998; Reinartz and Gruppe, 1993). 'Cooperation' in this sense can be interpreted widely, from perceptual support (e.g., Head-Up Display (HUDs)) to fully automating driving subtasks. Obviously, the cooperation issues at stake in car driving concern intermediary cooperation modes between the fully manual mode and fully automatic modes, as stressed in recent approaches developed by Harms (2006), or Young and Stanton (2006). The automation, then, is acting as any other team member in a multi-agent system, by offering help and/or advice as appropriate.

Any good team is built up of members with complementary skills, such that the task demands can be met and the team's goals achieved. If any of the team members happen to be machine, rather than human, the situation should be no different. Schutte (1999) coined the term 'complemation' to sum up the principle of exploiting automation to enhance human capabilities while compensating for their limitations. For instance, we could allow the human to provide creativity and adaptability, while letting the machine store information and make precise measurements—maximising the strengths of each team member.

Hoc (2001) and Hoc, Young and Blosseville (2009) explicitly considered human–machine cooperation to propose a more teamwork-based allocation of function model. The model proposes three levels of cooperation and four modes of task automation in a matrix of human–automation interaction. The crucial distinction is the establishment of a 'common frame of reference'—essentially a shared mental model of system operation, which must be held by both human and machine about the other's behaviour. One of the key aspects of this is the awareness of context and intent—vital in interpreting the actions implemented in a given situation. Hoc, Young and Blosseville (2009) provide the example of a lane-keeping system in cars—which warns the driver when straying out of his/her lane. Clearly, sometimes this activity is legitimate—when overtaking, for example. The system has access to the vehicle's electronics, though, and so only provides a warning if the driver is moving across the lane markings without having used a turn signal. Whilst this is a crude rule, it illustrates the importance of intent and context in maintaining that common frame of reference and hence the smooth dynamics of the team. In an abnormal scenario, for instance, the human needs support from the automation in order to resolve the situation. The problem now is that most automated systems are not designed to behave like humans when it comes to teamwork.

Dekker (2004) makes the similar point that it is not the quantity of automation which causes the problem, but the quality. Instead of designing automation on a 'who does what' basis, successful automation depends on designers answering the question of 'how do we get along?' On the flight deck, such teamwork comes under the umbrella of crew resource management (CRM) (Wiener, Kanki and Helmreich, 1993). Rather than thinking of CRM purely in terms of liveware-to-liveware or liveware-to-software (cf. Fitzgerald, 1997), though, why not also invoke it for software-to-liveware scenarios? After all, as Jensen (1997) points out, CRM is all very well, but the first line of defence should be the design of the system, not the flightcrew.

Young, Stanton and Harris (2007) argue for taking the concept a step further, applying the principles of CRM training to the design of automated systems. CRM-designed automation can be achieved in two ways. Firstly, by erring on the side of 'soft' automation, thus leaving the human in active control and able to delegate tasks as appropriate—in line with the frameworks proposed by Parasuraman, Sheridan and Wickens (2000) and Hoc (2001). Secondly, the teamworking aspect ultimately comes down to communication in both directions—which means a significant design effort on the control–display interface to optimise the flow of information (cf. Griffin, Young and Stanton, 2010).

If we are expecting the automation to behave as a team member—coordinating and cooperating with the driver—then we should apply notions of team performance to automation design. That means effective communications, group processes, team decision making, leadership, shared situation awareness, conflict resolution, and recognition of others' behavioural styles (Jensen, 1997). Christoffersen and Woods (2000) suggest that the more powerful automated systems become (i.e., high in autonomy and authority), the more feedback they need to supply to make their behaviour observable. From the human's perspective, if the capabilities and activities of automation are more transparent (as would be expected of a human colleague), then the problems of situation awareness, workload and trust should be ameliorated. Indeed, models of trust in automation have been built upon models of interpersonal trust in humans (Muir, 1994), so it seems logical to apply another aspect of human–human cooperation to human–machine cooperation.

In designing optimal human–human teams, the aims are to have a balance of skills and good communication and understanding between team members. Likewise for human–machine teams, the ultimate objective would be to design an automated system with complementary taskwork skills (cf. Schutte, 1999) and good teamworking abilities (cf. Hoc, 2001). As far as teamwork is concerned, though, the technological barriers are much higher. Although technology undoubtedly has some way to go before automation can be smart enough to know what the human wants to do, very recent technological breakthroughs suggest some potential solutions may be on the horizon.

Haufe et al. (2011) took a neuroergonomics approach to emergency braking in a driving simulator study, and managed to detect the driver's intention to brake via muscle and brain activity fractions of a second earlier than via actual pedal movements. If the appropriate sensors could feasibly be installed in cars, this could be a prime source of data for the technology to know what the driver is thinking—to sense his or her intent, and to match its actions accordingly. Similarly, the eye-tracking systems described earlier in this chapter could, with appropriate algorithms, make a reasonable guess at the driver's attention patterns, knowing where they are looking and, perhaps, what they are attending to.

Human-centred automation thus enables the driver and the car to work together as a team, optimising performance and—crucially in the present context—designing the technology to adapt to the driver, rather than the other way around. Applying the team–worker approach to automation design, not only should the technology be designed to support those driving tasks, which are more demanding or more susceptible to error, it should also be programmed to genuinely work with the human in charge of the vehicle. But it is not enough simply to actively cooperate based on behavioural data—the automation must also communicate its actions and intentions back to the driver in a clear and timely fashion. Such feedback has to be transparent enough that the driver knows not just what the system is doing, but is also aware of its design limitations. It cannot be emphasised enough just how crucial the role of communication is in maintaining the human–automation relationship. Ultimately, this all comes back to interface design.

5.4 ERGONOMICALLY DESIGNED ADAS

Until such a time when automation is capable of fully assuming the driving task without any human input or supervision, it is essential to consider the ergonomic implications of ADAS design. The ergonomic approach to any design challenge is to design the system around the needs, capabilities and limitations of the user—in this case, the driver. Ergonomics researchers are not going to stop the tide of vehicle technology—nor, indeed, would we probably want to. But whilst we should not be shying away from technology, we should also be showing some restraint to ensure it is implemented in the right manner—the overriding philosophy being to support, rather than replace, the driver with technology (cf. Hoc, Young and Blosseville, 2009; Young, Stanton and Harris, 2007). If our smart cars are going to produce smarter drivers, ergonomics must sit in the middle.

This chapter has reviewed two philosophical stances to automation and has applied them to current and near-future vehicle technology systems. Two classifications of such technology have been derived—vehicle automation, in which low-level vehicle control aspects are automated, and driving automation, in which the driver is relieved of higher-level tactical or strategic tasks. Meanwhile, drawing on trends in aviation automation, we can also distinguish between hard protection (where the computer has ultimate authority) and soft protection (in which the human can override the automation if needs be).

Whilst we have noted that there is no clear distinction between hard and soft protection on the one hand, and vehicle and driving automation on the other, when focusing within driving automation it does seem that hard protection presents more problems than promises. Issues of trust, situation awareness and mental models are all apparent, whereas soft protection is largely associated with potential mental workload concerns. This broadly agrees with the general opinion towards technological support systems rather than automated replacement of the driver. Moreover, the soft automation philosophy of allowing the driver to choose how and when the system works, as well as being able to override it, is much more aligned to a socio-technical systems perspective, whereby humans and technology cooperate as a team to achieve an overall goal. Therefore, it seems that on the face of it, soft automation fits much better with ergonomics principles and research than hard automation.

Of course, the reality is not going to be as simple as that, with soft automation causing problems of mental workload, which can be equally detrimental to performance. In all likelihood, we will have to match different elements of the driving task with different philosophies for optimum performance. In a sense, we have already begun, with traditional vehicle automation mostly falling into the hard automation category, whilst driving automation systems straddle the boundaries of soft and hard. Rather than an overarching philosophy of soft or hard automation for driving (as has been seen in aviation), a blend throughout the driving subtasks may prove most efficient.

Classical models of automation design have taken a coarse approach, dividing functions between human and machine depending on which is better at performing them, or concentrating on levels of automation as a way of managing interaction with the human operator (e.g., Kaber and Endsley, 2004). However, as technology moves on apace, these divisions are becoming blurred, and more recent models see the human and machine as part of a team—thus putting the design focus on communication and cooperation.

The emphasis with this philosophy is on levels of cooperation rather than automation, being concerned with the level of authority that the automation possesses, and its ability to communicate with the human operator. One of the perennial problems with automation is (lack of) feedback—not telling the human operator of its activities creates an additional workload burden and potential of error for the driver. Likewise, it is crucial that if an automated system is to operate effectively, it should be aware of the task context—both in terms of the environment around the vehicle and of the driver's intentions. Thus, a good analogy in terms of automation design is to consider it in the same way as a human co-driver.

Ideally, any technological support system should act like a co-driver in the passenger seat—subtle enough so as not to cause interference, but accessible enough so as to provide assistance when needed. The idea of communication and cooperation between human and machine is very much at the forefront of contemporary thinking regarding the human factors of automated systems (e.g., Dekker, 2004; Hoc, Young and Blosseville, 2009; Schutte, 1999; Young, Stanton and Harris, 2007). Human and automation should work as a team—with all the principles and caveats of human–human teamworking.

If we are expecting the automation to behave as a team member—coordinating and cooperating with the driver—then perhaps we should apply notions of team performance to automation design. An exemplar would be to adopt the principles of crew resource management (CRM) on flight decks (Wiener, Kanki and Helmreich, 1993). CRM focuses on communication, teamwork, and a two-way management structure that allows junior members of the flight crew to provide credible input into the decision-making process. As a set of principles for human–human cooperation, it has worked very well—so perhaps it would work equally well as a set of design principles for human–machine cooperation.

Treating the automation as a team member may put a different perspective on the attitude to design, but it does epitomise the human-centred design approach. Human-centred automation enables the driver and the car to work together as a team, optimising performance and—crucially in the present context—designing the technology to adapt to the driver, rather than the other way around. Adaptive automation—if it can be properly achieved—truly fits the equipment to the user.

A number of adaptive systems have already been discussed in the opening section to this chapter. Adaptive systems possess a level of intelligence so that they can change the level of information or support offered to the driver or alter system thresholds and parameters in real-time. Typically, the system will use sensors to detect some parameter of the task context, and will infer the driver's state based upon this information. The interface itself then adapts by providing more or less information depending upon the data that have been collected.

The concept of adaptive interfaces has been around for some time; historically, such systems have largely been discussed in relation to regulating mental workload, in order to maintain an optimal state for the operator (e.g., Byrne and Parasuraman, 1986; Hancock and Verwey, 1997). An adaptive system monitors the task situation for peaks of workload and, in such cases, automatically relieves the operator of some elements of the task. These tasks are returned to the operator when demand returns to a more manageable level. Workload regulation on the part of an adaptive system could be based on a previously stored model of the driver and/or the task context (Verwey, 1993), or in response to measures taken during the task. Such measures would have to be dynamically sensitive to workload, so that the system can continually and rapidly adapt to changes in demand, and could again comprise metrics of the driver, the vehicle, or the environment.

Early investigations of systems for real-time adaptation to workload pursued physiological metrics for continuous driver monitoring (Fairclough, 1993; Kramer, Trejo and Humphrey, 1996). Thus, one could imagine heart rate or respiration sensors in the driver's seat, or electrodermal response receptors in the steering wheel. Advances in technology and sensors have meant that more recent systems have moved away from physiological measurement as such, in favour of more overt behavioural indices or stored models of the driver. In the European AIDE project (see e.g., Amditis et al., 2010; Engström and Victor, 2009), sensors monitored the driver–vehicle environment system, using eye and head tracking, on-board diagnostics, and GPS map data respectively. These data were compared against a stored model of driver workload for a range of scenarios, as defined by experts and empirical testing (e.g., Tango et al., 2010).

This is the kind of approach taken by Piechulla et al. (2003) with their prototype adaptive interface for driver workload. Situational factors were detected by an on-board geographical database, and a computational workload estimator compared these data to a complex task-based model in order to assess those situations. Such situations included road type (urban, rural etc.), curvature, slope, junctions, and directions. If workload was deemed to have exceeded a set threshold, then incoming telephone calls were routed directly to voicemail without informing the driver. This adaptive interface showed promising results in terms of managing driver mental workload. Similar systems have recently been offered in cars from Volvo and Saab, which estimate driver workload from driver inputs (steering, acceleration, braking) in order to reschedule emails and phone calls (Engström and Victor, 2009).

Whilst most adaptive systems use real-time monitoring as their basis for adaptation, Young, Birrell and Davidsson (2011) offered a kind of 'temporal adaptation' as a means to overcome longer-term peaks and troughs in workload. The assumption is that drivers have spare capacity during low-workload periods of driving (e.g., motorway cruising) which could usefully be occupied by presenting advance information about task-relevant activities in the near future. For instance, rather than providing satellite navigation instructions at a time when the driver is already under high demands (i.e., at the junction), such information could be provided well in advance when workload is much lower. Although this kind of temporal adaptation did not improve objective performance measures in a simulator study, it did increase subjective mental workload for the low-workload period of driving, suggesting some potential for cancelling out these peaks and troughs as anticipated.

Workload managers such as those described above are largely preventative systems—designed to anticipate and predict instances of overload or underload as a kind of 'primary protection' measure. Distraction can also be mitigated in real-time akin to secondary protection, by trying to redirect drivers' attention to the relevant roadway scene. Whereas Young, Birrell and Davidsson (2011) attempted to achieve this mitigation in advance with their temporal adaptation, real-time adaptive systems can use eye- or head-tracking to detect whether the driver is distracted and, if so, provide them with a warning. Some of these systems are designed to alert the driver purely to their own distraction, in an effort to 'train' the driver to be aware of potential distractions and thus adapt their behaviour (e.g., Donmez, Boyle and Lee, 2007; Engström and Victor, 2009).

Whilst many of these systems might seem like 'tomorrow's world' technology at the time of writing, developments in ADAS are progressing with seemingly exponential speed. Indeed, it has traditionally been the case that such systems are fitted to prestige models and marques, only later to filter down to mass-market vehicles. However, in 2011 Ford released its new Focus model (a mass-market offering), including several ADAS devices as a package option. We can be sure that during the shelf life of this book, much of what has been discussed in this chapter will become a reality. What we do not know is how well such systems will have been designed to integrate with the driver. The technological options available in today's cars offer many potential benefits, but also many challenges for researchers and practitioners in ergonomics. Society has also changed, such that drivers are at least as concerned about the environment and fuel economy as they are about road safety, while an ageing population is being increasingly reflected in the demographics of driving licence holders. ADAS technologies can potentially keep us mobile, safe, and can improve efficiency—but only if they work with the driver.

REFERENCES

Amditis, A., Pagle, K., Joshi, S. and Bekiaris, E. (2010). Driver-vehicle-environment monitoring for on-board driver support systems: Lessons learned from design and implementation. *Applied Ergonomics*, 41(2), 225–235.

Angell, L., Auflick, J., Austria, P., Kochhar, D., Tijerina, L., Biever, W., Diptiman, T., Hogsett, J. and Kiger, S. (2006). *Driver Workload Metrics Project: Task 2 Final Report. Crash Avoidance Metrics Partnership (CAMP)*. Farmington Hills, MI: DOT HS 810 635.

Bainbridge, L. (1982). Ironies of automation. In G. Johannsen et al. (eds.), *Analysis, Design and Evaluation of Man-Machine Systems.* pp. 151–157. New York: Pergamon.

Bar-Gera, H. and Shinar, D. (2005). The tendency of drivers to pass other vehicles. *Transportation Research Part F: Traffic Psychology and Behaviour*, 8, 429–439.

Beaty, D. (1995). *The Naked Pilot: The Human Factor in Aircraft Accidents.* Shrewsbury: Airlife Publishing Ltd.

Billings, C. (1993). *Aviation Automation.* New Jersey: Lawrence Erlbaum.

Bradley, M., Keith, S. and Wicks, C. (2007). *Car Technology—Time for an IQ Test?* Lansdown Lecture, May 9, London, England.

Broström, R., Engström, J., Agnvall, A. and Markkula, G. (2006). Towards the next generation intelligent driver information system (IDIS): The Volvo car interaction manager concept. *Proceedings of the 13th ITS World Congress*, London, 8–12 October 2006.

Buchholz, K. (2003). Driver Advocate for Chrysler. *Automotive Engineering International*, October, p. 42.

Byrne, E. A. and Parasuraman, R. (1996). Psychophysiology and adaptive automation. *Biological Psychology*, 42, 249–268.

Cacciabue, P. C. and Saad, F. (2008). Behavioural adaptations to driver support systems: A modelling and road safety perspective. *Cognition, Technology and Work*, 10(1), 31–39.

Christoffersen, K. and Woods, D. D. (2000). *How to Make Automated Systems Team Players.* Columbus, OH: Institute for Ergonomics, The Ohio State University.

Dekker, S. (2004). On the other side of promise: What should we automate today? In D. Harris (ed.), *Human Factors for Civil Flight Deck Design.* pp. 183–198. Aldershot: Ashgate.

de Waard, D., Van der Hulst, M., Hoedemaeker, M. and Brookhuis, K. A. (1999). Driver behavior in an emergency situation in the automated highway system. *Transportation Human Factors*, 1, 67–82.

Donmez, B., Boyle, L. and Lee, J. (2007). Safety implications of providing real-time feedback to distracted drivers. *Accident Analysis and Prevention*, 39, 581–590.

Endsley, M. R. (1995). Toward a theory of situation awareness in dynamic systems. *Human Factors*, 37(1), 32–64.

Endsley, M. R. and Kiris, E. O. (1995). The out-of-the-loop performance problem and level of control in automation. *Human Factors*, 37(2), 381–394.

Engström, J. and Victor, T. W. (2009). Real-time distraction countermeasures. In M. A. Regan, J. D. Lee and K. L. Young (eds.), *Driver Distraction: Theory, Effects, and Mitigation.* pp. 465–483. Boca Raton, FL: CRC Press.

Ericsson, E., Larsson, H. and Brundell-Freij, K. (2006). Optimizing route choice for lowest fuel consumption—Potential effects of a new driver support tool, *Transportation Research Part C*, 14, 369–383.

Fairclough, S. (1993). Psychophysiological measures of workload and stress. In A. M. Parkes and S. Franzen (eds.), *Driving Future Vehicles.* pp. 377–390. London: Taylor & Francis.

Fitzgerald, R. E. (1997). Call to action: We need a new safety engineering discipline. *Professional Safety*, 42(6), 41–44.

Griffin, T. G. C., Young, M. S. and Stanton, N. A. (2010). Investigating accident causation through information network modelling. *Ergonomics,* 53(2), 198–210.

Haddad, H. and Musselwhite, C. (2007). *Prolonging Safe Driving Through Technology. SPARC Research Briefing Sheet 024*. Bristol: Centre for Transport and Society, University of the West of England.

Hancock, P. A., and Parasuraman, R. (1992). Human factors and safety in the design of intelligent vehicle–highway Systems (IVHS). *Journal of Safety Research*, 23(4), 181–198.

Hancock, P. A. and Verwey, W. B. (1997). Fatigue, workload and adaptive driver systems. *Accident Analysis and Prevention*, 29(4), 495–506.

Harbluk, J., Noy, Y., Trbovich, P. and Eizenman, M. (2007). An on-road assessment of cognitive distraction: Impacts on drivers' visual behaviour and braking performance. *Accident Analysis and Prevention*, 39, 372–379.

Harms, L. (2006). A Theoretical Analysis of Driver Support and Assistance Systems. *IEA Congress*, Maastricht, the Netherlands, July 9–14.

Harvey, R., Fraser, D., Bonner, D., Warnes, A., Warrington, E. and Rossor, M. (1995). Dementia and driving: Results of a semi-realistic simulator study. *International Journal of Geriatric Psychiatry*, 10, 859–864.

Haufe, S., Treder, M. S., Gugler, M. F., Sagebaum, M., Curio, G. and Blankertz, B. (2011). EEG potentials predict upcoming emergency brakings during simulated driving. *Journal of Neural Engineering*, 8, 056001 Stacks.iop.org/JNE/8/056001(accessed 8 May 2012).

Hoc, J. M. (2001). Towards a cognitive approach to human–machine cooperation in dynamic situations. *International Journal of Human–Computer Studies*, 54, 509–540.

Hoc, J. M. and Blosseville, J. M. (2003). Cooperation between drivers and in-car automatic driving assistance. In G. C. Van der Veer and J. F. Hoorn (eds.), *Proceedings of CSAPC'03*. pp. 17–22. Rocquencourt, France: EACE.

Hoc, J. M. and Lemoine, M. P. (1998). Cognitive evaluation of human–human and human–machine cooperation modes in air traffic control. *International Journal of Aviation Psychology*, 8, 1–32.

Hoc, J. M., Young, M. S. and Blosseville, J. M. (2009). Cooperation between drivers and automation: Implications for safety. *Theoretical Issues in Ergonomics Science*, 10(2), 135–160.

Hogema, J. H., Van Arem, B., Smulders, S. A., and Coëmet, M. J. (1997). Modelling changes in driver behaviour: on the effects of autonomous intelligent cruise control. In T. Rothengatter and E. Carbonell Vaya (eds.), *Traffic and Transport Psychology: Theory and Application*. pp. 237–246. Oxford: Pergamon.

Horberry, T., J., Anderson, J., Regan, M., Triggs, T. and Brown, J. (2006). Driver distraction: The effects of concurrent in-vehicle tasks, road environment complexity and age on driving performance. *Accident Analysis and Prevention*, 38, 185–191.

Hughes, P. K. and Cole, B. L. (1986). What attracts attention when driving? *Ergonomics* 29, 311–391.

Hughes, D., and Dornheim, M. A. (1995). Accidents direct focus on cockpit automation. *Aviation Week and Space Technology*, January 30, 52–54.

Jensen, R. S. (1997). The boundaries of aviation psychology, human factors, aeronautical decision making, situation awareness, and crew resource management. *International Journal of Aviation Psychology*, 7(4), 259–267.

Kaber, D. B., and Endsley, M. R. (1997). Out-of-the-loop performance problems and the use of intermediate levels of automation for improved control system functioning and safety. *Process Safety Progress*, 16(3), 126–131.

Kaber, D. B. and Endsley, M. (2004). The effects of level of automation and adaptive automation on human performance, situation awareness and workload in a dynamic control task. *Theoretical Issues in Ergonomics Science*, 5, 113–153.

Kantowitz, B. H., and Campbell, J. L. (1996). Pilot workload and flightdeck automation. In R. Parasuraman and M. Mouloua (eds.), *Automation and Human Performance: Theory and Applications*. pp. 117–136. Mahwah, NJ: Lawrence Erlbaum Associates.

Keith, S., Bradley, M., Wilson, J. and Whitney, G. (2007). The development of a participatory research methodology with older drivers. *Proceedings of the TRANSED Conference.* Montreal, Canada, June, 18–21.

Kramer, A. F., Trejo, L. J., and Humphrey, D. G. (1996). Psychophysiological measures of workload: Potential applications to adaptively automated systems. In R. Parasuraman and M. Mouloua (eds.), *Automation and Human Performance: Theory and Applications.* pp. 137–162. Mahwah, NJ: Lawrence Erlbaum Associates.

Labiale, G. (1997). Cognitive ergonomics and intelligent systems in the automobile. In Y. I. Noy (ed.), *Ergonomics and Safety of Intelligent Driver Interfaces.* pp. 169–184. Mahwah, NJ: Lawrence Erlbaum Associates.

Landau, K. (2002). The development of driver assistance systems following usability criteria. *Behaviour & Information Technology*, 21(5), 341–344.

Lee, J. D., and Moray, N. (1994). Trust, self-confidence, and operators' adaptation to automation. *International Journal of Human–Computer Studies*, 40, 153–184.

Lees, M. N. and Lee, J. D. (2007). The influence of distraction and driving context on driver response to imperfect collision warning systems. *Ergonomics*, 50(8), 1264–1286.

Liu, B. S. and Lee, Y. H. (2006). In-vehicle workload assessment: Effects of traffic situations and cellular telephone use. *Journal of Safety Research*, 37, 99–105.

Lundberg, C. (2003). *Older Drivers with Cognitive Impairments: Issues of Detection and Assessment.* Stockholm, Sweden, Karolinska Institutet.

Ma, R. and Kaber, D. B. (2005). Situation awareness and workload in driving while using adaptive cruise control and a cell phone. *International Journal of Industrial Ergonomics*, 35, 939–953.

May, A., Ross, T. and Osman, Z. (2005). The design of next generation in-vehicle navigation systems for the older driver. *Interacting with Computers*, 17, 643–659.

Molloy, R., and Parasuraman, R. (1996). Monitoring an automated system for a single failure: Vigilance and task complexity effects. *Human Factors*, 38(2), 311–322.

Muir, B. M. (1994). Trust in automation: Part I. Theoretical issues in the study of trust and human intervention in automated systems. *Ergonomics*, 37(11), 1905–1922.

Muir, B. M., and Moray, N. (1996). Trust in automation: Part II. Experimental studies of trust and human intervention in a process control simulation. *Ergonomics*, 39(3), 429–460.

Nilsson, L. (1995). Safety effects of adaptive cruise control in critical traffic situations. In *Proceedings of the second world congress on intelligent transport systems: Vol. 3.* pp. 1254–1259. Tokyo: Vehicle, Road and Traffic Intelligence Society.

Norman, D. A. (1990). The 'problem' with automation: Inappropriate feedback and interaction, not 'over-automation'. *Phil. Trans. R. Soc. London B*, 327, 585–593.

Palmer, E. (1995). 'Oops, it didn't arm.'—A case study of two automation surprises. *8th International Symposium on Aviation Psychology.* Columbus, Ohio: Ohio State University.

Parasuraman, R. (2000). Designing automation for human use: Empirical studies and quantitative models. *Ergonomics*, 43(7), 931–951.

Parasuraman, R., Mouloua, M., Molloy, R., and Hilburn, B. (1996). Monitoring of automated systems. In R. Parasuraman and M. Mouloua (eds.), *Automation and Human Performance: Theory and Applications.* pp. 91–115. Mahwah, NJ: Lawrence Erlbaum Associates.

Parasuraman, R. and Riley, V. (1997). Humans and automation: Use, misuse, disuse, abuse. *Human Factors*, 39(2), 230–253.

Parasuraman, R., Sheridan, T. B. and Wickens, C. D. (2000). A model for types and levels of human interaction with automation. *IEEE Transactions on Systems, Man, and Cybernetics—Part A: Systems and Humans*, 30(3), 286–297.

Piechulla, W., Mayser, C., Gehrke, H. and König, W. (2003). Reducing drivers' mental workload by means of an adaptive man–machine interface. *Transportation Research Part F*, 6, 233–248.

Ranney, T. A. (1994). Models of driving behavior: A review of their evolution. *Accident analysis and prevention*, 26(6), 733–750.

Regan, M. A., Lee, J. D. and Young, K. L., eds. (2009). *Driver Distraction: Theory, Effects, and Mitigation.* Boca Raton, FL: CRC Press.

Reinartz, S. J. and Gruppe, T. R. (1993). Information requirements to support operator-automatic cooperation. *Human Factors in Nuclear Safety Conference,* IBC, London, April 22–23.

Richardson, M., Barber, P., King, P., Hoare, E., and Cooper, D. (1997). Longitudinal driver support systems. *Proceedings of Autotech 1997.* pp. 87–97. London: IMechE.

Rudin-Brown, C. M. and Parker, H. A. (2004). Behavioural adaptation to adaptive cruise control (ACC): Implications for preventive strategies. *Transportation Research Part F: Traffic Psychology and Behaviour,* 7(2), *59–76.*

Sarter, N. B., and Woods, D. D. (1995). How in the world did we ever get into that mode? Mode error and awareness in supervisory control. *Human Factors,* 37(1), 5–19.

Schutte, P. (1999). Complemation: An alternative to automation. *Journal of Information Technology Impact,* 1(3), 113–118.

Shinar, D., Meir, M. and Ben-Shoham, I. (1998). How automatic is manual gear shifting? *Human Factors,* 40(4), 647–654.

Smith, M. R. H., Witt, G. J., Bakowski, D. L., Leblanc, D. and Lee, J. D. (2009). Adapting collision warnings to real-time estimates of driver distraction. In M. A. Regan, J. D. Lee, and K. L. Young (eds.), *Driver Distraction: Theory, Effects, and Mitigation.* pp. 501–518. Boca Raton, FL: CRC Press

Stanton, N. A. and Marsden, P. (1996). From fly-by-wire to drive-by-wire: Safety implications of automation in vehicles. *Safety Science,* 24(1), 35–49.

Stanton, N. A. and Young, M. S. (2000). A proposed psychological model of driving automation. *Theoretical Issues in Ergonomics Science,* 1(4), 315–331.

Stanton, N. A. and Young, M. S. (2005). Driver behaviour with adaptive cruise control. *Ergonomics,* 48, 1294–1313.

Stanton, N. A., Young, M. and McCaulder, B. (1997). Drive-by-wire: The case of driver workload and reclaiming control with adaptive cruise control. *Safety Science,* 27(2/3), 149–159.

Taieb-Maimon, M. and Shinar, D. (2001). Minimum and comfortable driving headways: Reality versus perception. *Human Factors,* 43, 159–172.

Tango, F., Minin, L., Tesauri, F. and Montanari, R. (2010). Field tests and machine learning approaches for refining algorithms and correlations of driver's model parameters. *Applied Ergonomics,* 41(2), 211–224.

Van der Voort, M., Dougherty, and Van Maarseveen, M. (2001). A prototype fuel efficiency support tool. *Transportation Research Part C,* 9, 279–296.

Van Driel, C., Hoedemaeker, M. and Van Arem, B. (2007). Impacts of a congestion assistant on driving behaviour and acceptance using a driving simulator. *Transportation Research Part F,* 10, 139–152.

Verwey, W. B. (1993). How can we prevent overload of the driver? In A. M. Parkes and S. Franzen (eds.), *Driving Future Vehicles.* pp. 235–244. London: Taylor & Francis.

Walker, G. H., Stanton, N. A., and Young, M. S. (2001). Where is computing driving cars? *International Journal of Human Computer Interaction,* 13(2), 203–229.

Ward, N. J. (2000). Task automation and skill development in a simplified driving task. In *Proceedings of the XIVth Triennial Congress of the International Ergonomics Association and 44th annual meeting of the Human Factors and Ergonomics Society,* Vol. 3. pp. 302–305. Santa Monica, CA: HFES.

Ward, N. J., Fairclough, S., and Humphreys, M. (1995). The effect of task automatisation in the automotive context: A field study of an autonomous intelligent cruise control system. *International Conference on Experimental Analysis and Measurement of Situation Awareness.* November 1, Daytona Beach, Florida.

Wickens, C. D., Gordon, S. E. and Liu, Y. (1998). *An Introduction to Human Factors Engineering*. New York: Longman.

Wiener, E. L., and Curry, R. E. (1980). Flight-deck automation: Promises and problems. *Ergonomics*, 23(10), 995–1011.

Wiener, E. L., Kanki, B. G. and Helmreich, R. L. (1993). *Cockpit Resource Management*. San Diego: Academic Press.

Young, M. S. and Birrell, S. A. (2012). Ecological IVIS Design: Using EID to develop a novel in-vehicle information system. *Theoretical Issues in Ergonomics Science*, 13(2), 225–239.

Young, M. S., Birrell, S. A. and Davidsson, S. (2011). Task pre-loading: Designing adaptive systems to counteract mental underload. In M. Anderson (ed.), *Contemporary Ergonomics and Human Factors*. pp. 168–175. London: Taylor & Francis.

Young, M. S. and Bunce, D. (2011). Driving into the sunset: Supporting cognitive functioning in older drivers. *Journal of Aging Research*, Vol. 2011, Article ID 918782, 6 pages. doi:10.4061/2011/918782.

Young, M. S. and Stanton, N. A. (2001). Size matters. The role of attentional capacity in explaining the effects of mental underload on performance. In D. Harris (ed.), *Engineering Psychology and Cognitive Ergonomics: Vol. 5—Aerospace and Transportation Systems*. Proceedings of the Third International Conference on Engineering Psychology and Cognitive Ergonomics, Edinburgh, October 25–27 2000. pp. 357–364. Aldershot: Ashgate.

Young, M. S. and Stanton, N. A. (2002). Malleable attentional resources theory: A new explanation for the effects of mental underload on performance. *Human Factors*, 44(3), 365–375.

Young, M. S. and Stanton, N. A. (2006). How do you like your automation? The merits of hard and soft in vehicle technology. *IEA Congress*, Maastricht, the Netherlands, 9–14 July 2006.

Young, M. S. and Stanton, N. A. (2007a). Back to the future: Brake reaction times for manual and automated vehicles. *Ergonomics*, 50(1), 46–58.

Young, M. S. and Stanton, N. A. (2007b). What's skill got to do with it? Vehicle automation and driver mental workload. *Ergonomics*, 50(8), 1324–1339.

Young, M. S., Stanton, N. A. and Harris, D. (2007). Driving automation: Learning from aviation about design philosophies. *International Journal of Vehicle Design*, 45(3), 323–338.

6 Human Response to Vehicle Vibration

Neil J. Mansfield
Loughborough University, UK

CONTENTS

6.1   INTRODUCTION

All vehicles expose their occupants to some form of vibration. Vibration can be due to the inherent motion of the vehicle, such as manoeuvring, due to in-vehicle sources such as motors, or due to the surface on which the vehicle is travelling. In most cases, the focus of vehicle ergonomics is on the driver, but in many cases, vehicles will contain multiple people, all of whom are exposed to vibration. This vibration is usually transmitted to the occupant through a seat, but could also be transmitted through contact with the hands, through the feet, or via headrests. Some travellers are exposed whilst standing, others whilst lying down. This chapter focuses on those exposed to vibration whilst sitting on vehicle seats.

At low magnitudes, vibration can be annoying or distracting, at higher magnitudes it can cause activity interference and discomfort. In extreme cases, vibration can be a health hazard and cause chronic or acute injury presenting as low back, neck and shoulder pain. Due to the potential hazards from vibration, a legal framework has been developed in Europe to protect professional drivers from vibration. This is defined as the Physical Agents (Vibration) Directive (European Commission 2002) and defines vibration limits, and a framework for action if risk is present. Most non-professional drivers are unlikely to approach health risk thresholds unless they drive for long periods of time, drive off-road or on poor road surfaces.

Optimisation of the response to the vibration environment to which a vehicle occupant is exposed can be achieved through either modifying the vibration or modifying the task, which they are required to perform. In addition to the design of the vehicle and the state of the terrain, the vibration can be affected by driver behaviour whereby a skilled driver can reduce the vibration by selecting speed and route appropriately; this is of particular importance in passenger transportation where the driver can occupy a better performing seat than their passengers, and therefore be better isolated. In this case, a smoother ride can be experienced by passengers through improved driving technique.

In general, vehicles with larger wheels and heavier bodies will vibrate less than those with small wheels and lightweight bodies when driving on the same surface. The best performing vibration isolation systems (suspensions and seats) are usually those that are heavier. With current trends requiring lighter-weight vehicles in order to reduce fuel consumption and emissions, the response of vehicles to road surface irregularities may become more of a challenge, and careful design, optimisation, and compromise become more important.

6.2   THE VIBRATION ENVIRONMENT IN VEHICLES

In order to understand the likely response of a person exposed to vibration, it is necessary to characterise the vibration including its direction, magnitude, frequency content and the duration of exposure. Whilst much can be learned from detailed measurements of an environment, and in many cases, such measurements are essential, care must be taken to understand the inherent variability in the vibration emission of a vehicle, such that any small improvements in performance can be put into context. Drivers' behaviour can change from day-to-day due to time pressures, mood or traffic. Road conditions can change with weather (e.g., snow), through deterioration or maintenance. Under ideal conditions on the same road surface, coefficients of variation of 5 to 8% have been reported for 60 repeat vibration measurements of the same car driving the same route at the same time of day (Paddan, 2004).

For automotive applications, vibration co-ordinate systems are defined as x-axis, y-axis and z-axis for fore-and-aft, lateral and vertical motion, respectively (Figure 6.1). Rotational motion around these axes are defined as roll, pitch and yaw in ISO 2631-1 (International Organization for Standardization (ISO), 1997). The origin of the co-ordinate systems for practical measurements of seated drivers is usually defined as the centre of the seat cushion. As seat cushions are rarely horizontal, the co-ordinate system is slightly rotated to be aligned with the interface of the seat and the body.

Vibration magnitude is usually reported in m/s^2 r.m.s. This is calculated from the acceleration measured at the measurement point. In order to model the response of the body to different frequencies of motion, the acceleration signal is frequency weighted using digital signal processing techniques. Mathematically, the r.m.s. can be expressed as:

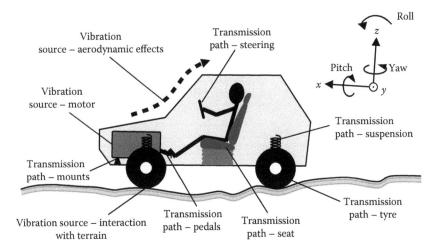

FIGURE 6.1 Vibration sources and vibration transmission paths for a vehicle travelling on a rough road. The figure also shows the whole-body vibration co-ordinate system for seat–surface vibration.

$$a_{w\,r.m.s.} = \sqrt{\frac{1}{T} \int_0^T a_w^2(t)\,dt} \qquad (6.1)$$

where

$a_{w\,r.m.s.}$ is the frequency-weighted r.m.s. acceleration,

T is the measurement duration and

$a_w(t)$ is the frequency-weighted acceleration at time t.

Typical vibration magnitudes in road vehicles driving on asphalt roads are between 0.2 and 0.6 m/s² r.m.s. in the most severe (usually vertical) direction (Paddan and Griffin, 2002a; Marjanen 2010). Off-road driving can expose drivers to magnitudes exceeding 1.5 m/s² r.m.s.

Vibration exposures are rarely continuous for any extended period of time as traffic flow, road surfaces and road types vary. It can therefore be helpful to combine exposures into a single dose measure. The most commonly used metrics are the A(8), the daily vibration exposure normalised to an 8-hour period, and the vibration dose value (VDV). To calculate A(8), vibration exposures are summed using the expression:

$$A(8) = \sqrt{\frac{1}{8} \sum_{n=1}^{n=N} a_{wn}^2 t_n} \qquad (6.2)$$

where

a_{wn} and t_n are the frequency-weighted r.m.s. acceleration and exposure time (in hours) for exposure n, and

N is the number of exposures (Figure 6.2).

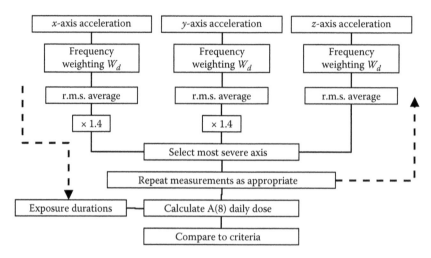

FIGURE 6.2 Illustration of the process of analysing tri-axial whole-body vibration data for the purposes of calculating the A(8).

If the individual being assessed is exposed for more or less than 8 hours, the sum of the individual exposures could be greater than or less than 8. Nevertheless, the 8-hour reference exposure time is retained (Mansfield, 2005). Mathematically, the VDV can be expressed as:

$$VDV = \sqrt[4]{\int_0^T a_w^4(t)\,dt} \qquad (6.3)$$

where

T is the measurement duration and

$a_w(t)$ is the frequency-weighted acceleration at time t.

Both the VDV and A(8) increase with increased vibration exposure and with increased vibration magnitude. As they are a function of measurement time, care must be taken in comparing measurements of different duration. According to the Physical Agents (Vibration) Directive, vibration exposures can be considered potentially hazardous to health if the A(8) exceeds 0.5 m/s² r.m.s. or the VDV exceeds 9.1 m/s¹·⁷⁵ in any 24-hour period, and risk management action must be taken.

Vibration experienced on the seats of cars is dominated by vibration at frequencies below about 30 Hz (Qiu and Griffin, 2003; Mansfield and Griffin, 2000; Griffin, 1978). The dynamics of the seat itself serves as a vibration isolator to high-frequency vibration in the vertical direction (see Section 6.4). At frequencies below about 1 Hz, vibration is caused by the interaction of the moving vehicle with undulations of the ground, and can be affected by aerodynamic buffeting. For vibration between 1 Hz and 20 Hz, the primary cause is road irregularities. At higher frequencies, vibration can be generated by the engine. These vibrations use usually more detectable at the hands and the feet than the seat, due to the seat cushion attenuation.

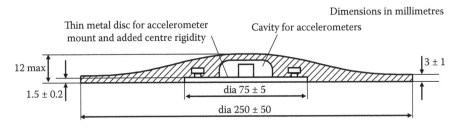

FIGURE 6.3 Cross section of design of flexible disc for mounting seat accelerometers as defined in ISO 10326-1 (ISO, 1992).

Measurements of vehicle vibration are made using accelerometers mounted to the vehicle as close to the contact point as possible. For measurements of vibration on a vehicle seat, accelerometers are mounted into a thin flexible disc and this is placed on the seat cushion (Figures 6.3 and 6.4). The bodyweight of the occupant sitting on the disc keeps the embedded accelerometers in contact with the seat and the flexible disc ensures an acceptable level of comfort. Care must be taken to avoid signals caused by the occupant, rather than the vehicle, such as seat ingress and egress, or shuffling in the seat, known as 'seat motion artefacts'. Signals from accelerometers are gathered using data acquisition hardware and signal processing software that allows for calibration, frequency weighting and calculation of vibration metrics and/ or vibration spectra to show the frequency content of the vibration. Tests must always be completed with a human occupant, as their body dynamics will change the vibration at the seat surface. An inert or 'crash test' dummy is not a suitable test seat occupant as their dynamic responses do not replicate those of humans.

FIGURE 6.4 Flexible disc containing accelerometers mounted on an automobile seat.

6.3 PERCEPTION OF AND DISCOMFORT FROM VIBRATION

The human body detects vibration through a variety of sensory systems including those in the vestibular (balance) organs, through several types of nerve endings in the skin, and through proprioception via soft tissues (Griffin, 1990). Each of these systems is more sensitive to some frequencies of vibration than to others. An integrated sensation of the environment will also be enhanced through other channels such as visual and auditory feedback, and this can make it difficult to accurately determine the relative disturbances felt from each system in a real vehicle as it is difficult for one modality to be considered in isolation (Mansfield, Ashley and Rimell, 2007).

Human response to vibration has been widely tested in the laboratory and has shown that perceptions of vibration are most sensitive at those frequencies where the body has its biomechanical resonances. In the vertical direction, the body resonates at about 5 Hz; in horizontal directions, the resonances occur below 2 Hz, but are highly dependent on the direction of the motion, the posture, and backrest contact. If subjective methods are used to establish the relative sensitivity to vibration of different frequencies, a similar conclusion can be drawn, that the body is most sensitive at 5 Hz vertically and 2 Hz horizontally.

The relative response of the body to vibration of different frequencies can be modelled using a frequency weighting. A frequency weighting is a multiplication factor that varies with frequency of the signal and is usually implemented in the measurement instrumentation. If the weighting has a low value, this indicates that the body is less responsive at that frequency; if the weighting has a relatively high value, this indicates that the body is more responsive at that frequency. The most commonly used frequency weightings are W_d for horizontal vibration at the seat and W_k for vertical vibration at the seat (Figure 6.5). Both of these weightings are defined in ISO 2631-1 (1997). In some cases, other weightings might be appropriate if assessing vibration of the floor, at the head, or rotational motion.

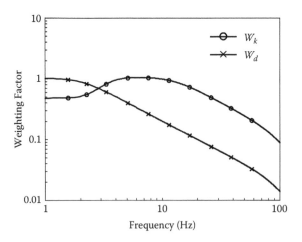

FIGURE 6.5 Frequency–response functions of frequency weightings W_d and W_k as defined in ISO 2631-1 (ISO, 1997).

Vehicle vibration occurs simultaneously in more than one direction at a time. ISO 2631-1 (1997) describes several possible methods for combining vibration from component directions of motion into a single metric in order to predict the discomfort. The simplest of these methods uses tri-axial measures of vibration at the seat cushion and scales the two horizontal axes by 1.4. These three vibration components are then combined using the 'root sum of squares' method to obtain a point vibration total value a_v:

$$a_v = \sqrt{\left(k_x^2 a_{wx}^2 + k_y^2 a_{wy}^2 + k_z^2 a_{wz}^2\right)} \tag{6.4}$$

where

> a_{wx}, a_{wy}, and a_{wz} are the weighted r.m.s. accelerations in the x- , y- , and z-axes respectively and k_x, k_y, and k_z are multiplying factors of 1.4 in x- , and y-axes and 1.0 in the z-axis.

It has been suggested that the 1.4 factors under-estimate the contribution of horizontal vibration, and better correlations between vibration and subjective responses are obtained with values greater than 2 (Greifahn and Bröde, 1999; Marjanen and Mansfield, 2010; Mansfield and Maeda, 2011).

Frequency-weighted vibration magnitude alone is insufficient to determine the overall discomfort in a vehicle seat, due to the contribution of many other factors. These factors include the seat design, seat stiffness, and thermal factors. Ebe and Griffin (2000a, 2000b, 2001) produced a conceptual model of this where they classified those non-vibration related aspects as 'static' factors, and those vibration related aspects as 'dynamic' factors. For situations where there is no vibration, the overall discomfort is dictated by the static factors alone. This represents times when the vehicle is not moving, such as in a showroom when a potential purchaser might be gaining their first impressions of the seat comfort. Once the car is being driven, there will be some vibration and depending on the nature of the road, the vehicle, and the driving style, this could be a high- or low-vibration magnitude. The relative importance of the dynamic factors of the seat compared to the static factors are a function of the vibration magnitude. As the vibration magnitude increases, the relative importance of the dynamic factors increases and therefore the seating dynamics become a more critical aspect of the seat design (Figure 6.6).

Ebe's model is a useful representation of automotive comfort at any particular time, but does not include temporal aspects. When a person occupies a seat for an extended period of time, their discomfort gradually increases. Therefore, the model was extended conceptually to include temporal factors by adding a third dimension of 'time' (Mansfield, 2005). As time increases, the discomfort increases (Figure 6.7a). However, experimental work has shown that the increase in the discomfort occurs more rapidly when there is vibration compared to when there is not (Mansfield, 2010). Therefore, the presence of vibration becomes relatively more important over long-term driving, and thus measures to minimize the vehicle occupant's vibration

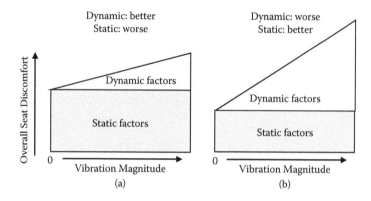

FIGURE 6.6 Ebe's model of overall seat discomfort showing the changing relative importance of static and dynamic parameters depending on the vibration magnitude. (Adapted from Ebe, K. and Griffin, M. J., 2000a, Qualitative Models of Seat Discomfort Including Static and Dynamic Factors, *Ergonomics*, 43(6), 771–790.)

exposure become more important. This acceleration of discomfort onset is termed 'dynamic fatigue' in Figure 6.7b.

Regression analysis has shown that any model designed to represent long-term overall discomfort when exposed to whole-body vibration needs to include factors able to represent the static discomfort (a constant for the seat), fatigue discomfort (a component which depends on time), vibration discomfort (a component which depends on the vibration magnitude), and dynamic fatigue (a component of interaction between the vibration exposure and duration). These four variables can be expressed as:

$$\Psi = s_s + f_t\,t + d_v\,a + i_{tv}\,ta \qquad\qquad (6.5)$$

where
Ψ is the rating of discomfort,
s_s is the static discomfort constant,
f_t is a fatigue constant,
d_v is the vibration discomfort constant,
i_{tv} is an interaction variable,
t is the time (mins) and
a is the frequency-weighted vibration.

Although this model is relatively simple, there is currently no evidence to indicate that a more complex model would generate a more representative result.

6.4 SEATING DYNAMICS

Automotive seats are capable of amplifying or attenuating the vibration to which the occupant could be exposed. Whilst each seat–occupant pairing will produce a slightly different dynamic response, there are general trends that are observed.

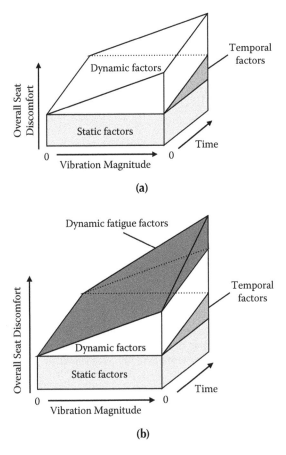

FIGURE 6.7 (a) Conceptual extension of Ebe's model of overall car seat discomfort showing static, dynamic and temporal factors. (b) Improved model of overall car seat discomfort including dynamic fatigue.

The dynamic response of a vehicle seat is a function of frequency. This frequency–response function is known as a measure of transmissibility and is defined as the ratio of the vibration on the surface of the seat (the 'output') to the vibration at the base of the seat (the 'input' at the floor) at any frequency,

$$T(f) = \frac{a_{seat}(f)}{a_{floor}(f)} \qquad (6.6)$$

where
   $T(f)$ is the transmissibility,
   $a_{seat}(f)$ is the acceleration on the seat and
   $a_{floor}(f)$ is the acceleration at the base of the seat at frequency f.

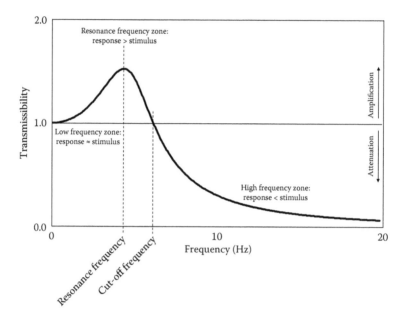

FIGURE 6.8 Typical vertical transmissibility measured for a conventional automotive seat. The seat provides vibration isolation at frequencies greater than the cut-off frequency and amplifies the vibration at frequencies lower than the cut-off frequency.

If there is the same magnitude of acceleration at the floor and on the seat surface, then the transmissibility is unity, as would be experienced if the seat were rigid. If the seat is providing isolation at some frequencies, then the transmissibility will be less than unity; at frequencies where the transmissibility is greater than unity, the seat will amplify the vibration. Transmissibilities are usually measured in the laboratory using a safety- and ethically- approved shaker.

At low frequencies, the seat transmissibility always tends to 1.0 (Figure 6.8). This is because for very slow oscillations, the seat moves as a single unit, and the response on the surface of the seat is similar to the stimulus at the base of the seat. If the transmissibility does not tend to 1.0, there could be a calibration error, or accelerometers on the seat surface and at the seat base might not be aligned with each other. As the frequency increases, there is relative movement between the seat surface and the seat base. The vibration energy can work in phase with the dynamics of the seat and motion can build up on the seat surface, such that the seat works as a vibration amplifier resulting in the response being greater than the stimulus. This amplification zone peaks at the resonance frequency, which occurs at about 4 Hz for most car seats. At frequencies greater than the resonance frequency, the transmissibility tends to reduce with increases in frequency. Above the cut-off frequency, the transmissibility is less than 1.0 and the seat works as a vibration isolator. A further complication is that the dynamics of automobile seats are non-linear, which means that the resonance frequency has a tendency to occur at a lower frequency for high-magnitude motion than for low-magnitude motion.

As the dynamic performance of a seat is a function of frequency, and the frequency of the vibration is a function of the instantaneous driving environment, the overall seat performance is not constant. For example, a seat will perform better at isolating vibration at 10 Hz compared to vibration at 5 Hz. If the road roughness changes in character from producing 10 Hz vibration to 5 Hz vibration, the seat will change from acting as an isolator to acting as an amplifier.

Where most car seats' dynamic performance is dictated by the characteristics of the foam from which it is constructed, many commercial-vehicle seats can also have a dedicated independent suspension designed to provide vibration isolation. These suspension seats comprise a mechanical isolator including some form of motion damping and either a pneumatic or coil spring. The seats are designed to reduce the frequency of the seat resonance, usually to about 2 Hz. This means that the 'high-frequency zone' where the response is less than the stimulus, extends over a greater frequency range and so less of the vibration is transmitted to the seat occupant. Suspension seats are relatively bulky and therefore only suitable for heavier vehicles where there is more scope for fitting a large seat. They are not found in cars, but common for bus and truck drivers, and a standard item in agricultural and mining mobile machinery. Some manufacturers have produced electronic suspension seats using active control technologies, but these tend to be costly and have not achieved a large market share.

An alternative method of characterising the vibration performance of a seat is to measure the ratio of the frequency weighted vibration on the seat surface to the frequency-weighted vibration at the base of the seat. This ratio is known as the 'seat effective amplitude transmissibility' or SEAT value (Griffin, 1990) (see Figure 6.9),

$$SEAT\% = 100 \times \frac{r.m.s._{.seat}}{r.m.s._{.floor}} \tag{6.7}$$

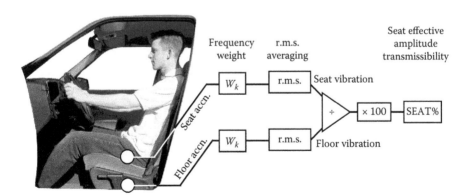

FIGURE 6.9 Graphical representation of the process for calculating the seat effective amplitude transmissibility (SEAT value) for a car seat using the W_k frequency weighting and the root mean square (r.m.s.) method.

where

r.m.s.$_{seat}$ is the frequency weighted vibration on the surface of the seat and
r.m.s.$_{floor}$ is the frequency weighted vibration at the base of the seat.

When the vehicle is travelling under conditions where the seat is providing overall isolation of the vibration, the SEAT value will be below 100%; if the seat is amplifying the overall vibration magnitude, then the SEAT value will be greater than 100%.

SEAT values can be used to compare different seats in the same vehicle under the same driving conditions, or comparing the effectiveness of a seat in a vehicle, which is used in a variety of driving conditions. The designer of the vehicle must prioritise under which conditions the SEAT value should be optimised.

For 25 cars, Paddan and Griffin (2002b) reported median SEAT values of 82%, with a range of 64 to 125%. Therefore, car seats usually reduce the vibration, but there are cases where seats are poorly matched to the vehicle, and amplification can occur. Using a modelling technique, the authors predicted the effect of all permutations of car and driver's seat on the vibration. It was shown that in all cases, at least one of the alternative seats would have improved the SEAT value and in one case, any of the other seats would have improved the vibration experienced in the car.

Understanding of the seat dynamics is an important tool in optimizing the driving environment. Whilst similar principals can be applied to any other part of the vehicle dynamic response, such as the tyres, wheel suspension, or, for some vehicles, the suspension of the cab, the biggest opportunity for improving the dynamic environment for the driver is usually through the design of the seat as other component parts of the vehicle are primarily designed for an alternative purpose.

6.5 PERFORMANCE UNDER VIBRATION

The very first automobiles in the 19th century were driven using several types of controls, rather than the steering wheel and pedals that we are now familiar with. As an evolution of the horse-drawn carriage, the 'horseless carriage' tended to keep the hands in the lap and was steered using a tiller, similar to steering a boat. Steering wheels were first developed as an innovation to improve the controllability for race vehicles of their period and started to be introduced to all automobiles around 1900. This change in primary control can be thought of as a form of natural selection whereby the less efficient tillers died out as the performance requirements increased and the pedal/steering wheel configuration proved more suitable.

In laboratory tests of steering and pedal performance for drivers exposed to vibration, there is usually very little degradation of performance until magnitudes become so great that they would not be experienced in normal driving. For example, even under deliberately compromised postural conditions, Baker and Mansfield (2009) showed no significant differences between steering performance measured under zero vibration and magnitudes similar to those experienced when driving off-road for a laboratory trial of a steering wheel task. One might initially interpret this as showing that vibration magnitude is not important for driver performance; however, if measures of workload using the standardized NASA-TLX test protocol (Hart and Staveland, 1988) are also considered, it is clear that drivers experience a greater

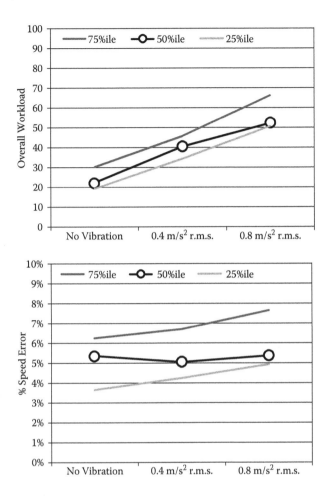

FIGURE 6.10 Driver workload and performance for a speed control task when exposed to vibration and using a driving simulator. There are no significant changes in workload with vibration, but significant increases in workload. (Data obtained from 12 subjects in a standard car driving posture.)

workload under vibration, as they need to adapt to the challenging physical environment to which they are exposed. Similarly speed control using a standard driving pedal setup shows no significant degradation in performance with vibration, whereas workload increases with vibration (Figure 6.10). Inferior controls, such as joysticks, can show a reduction in operator accuracy at magnitudes that could be experienced under normal driving, but again more impact on driver workload is observed than impact on objective measures of performance (Newell and Mansfield, 2008).

Whilst primary controls are well established and common for road vehicles (excluding motorcycles), secondary controls show more variation from control to control and between manufacturers. Increasing complexity of in-car information and entertainment systems has resulted in some systems that could be difficult to use under vibration.

A button, switch or position on a touch screen can be considered a target that a driver might need to interact with in order to complete a task. This target acquisition includes elements of visual perception and pointing tasks. Visual perception is most affected when a display moves relative to the observer, and least affected when the observer and display move in-phase (Moseley and Griffin, 1986). In an automobile, the target and the driver are both moving but the relative motion between them depends on frequency. At resonance, there can be substantial relative motion of the driver and their controls, due to the effects of the seat transmissibility, and due to the motion not occurring in-phase. This means that there is a slight delay between the motion at the base of the seat and the vibration reaching the occupant.

For pointing tasks, there is evidence to show that there can be more than twice as much motion at the hand than at the seat surface (Griffin, 1990). The peak in the seat to pointing-hand transmissibility occurs between 4 and 10 Hz and has a magnitude of between 2 and 3. This coincides with the peak in the floor to seat surface transmissibility which can also reach a magnitude of 2.0. In a worst case, it would be possible for these two resonances to interact creating a complex dynamic system such that vibration at the seat resonance frequency would cause relative motion between the hand and any target on a dashboard or in-vehicle device multiples of the vibration entering the base of the seat. This therefore makes it difficult for the driver to accurately select a target with a pointed finger.

An effective method improving the performance of a driver in the ability to select a small target is to ground the hand by allowing it to rest on the surround. This short-circuits the dynamics of the seat cushion and the dynamics of the extended arm and allows for more precision in target acquisition. Whilst this is usually possible for discrete controls fixed into a traditional dashboard, it is difficult to achieve for a large touch-screen where the entire area surrounding the target can be touch sensitive and operate unwanted controls. Fine motor control is easily degraded by vibration, and this degradation increases with vibration magnitude (e.g., Baker and Mansfield, 2010). It is usually better to avoid fine controls in vehicles, if the interface is to be operated whilst the vehicle is moving. Similarly, if the control is to be used by the driver whilst driving, then the eyes-off-the-road time should be minimized by ensuring that the interface can be easily located and operated whilst under dynamic conditions. If user testing is conducted without considering the vibration, then an important factor that could degrade performance may be ignored.

6.6 MOTION SICKNESS

Passengers in vehicles frequently report symptoms of motion sickness. These symptoms include bodily warmth, dry mouth, headache, drowsiness, lethargy, 'stomach awareness', nausea, pallour, and, the most tangible symptom, vomiting. There may be a lag from onset of motion to onset of motion sickness symptoms, but once they start to occur, the sequence of symptoms can be experienced rapidly.

The generally accepted theory to describe the onset of motion sickness is known as 'sensory rearrangement theory', or sometimes 'sensory conflict theory'. The basis of this theory is that one sensory system generates signals to the brain that are not in agreement with another set of sensory signals. Motion sensing parts of the body comprise the somatic system, vestibular system, visual system and 'control' system. If these give signals that differ to the expected combination of sensory signals then sensory conflict is said to have occurred.

Mansfield (2005) gave an example of a driver and passenger travelling in a car and turning a corner. The driver drives normally along the tree-lined road; the passenger has chosen to read during the journey. For the driver, the somatic system would sense changes in pressure internally and in contact with the seat at the turn at the corner. The vestibular system (balance organs located in the inner ear) would directly sense accelerations related to the forces due to cornering through motion of fluid in the semicircular canals and the otoliths. The visual system would show that when the car started to take the corner, objects would move across the visual field in combination with the optic flow of the passing objects. The driver being in control of the vehicle would know that they had turned the steering wheel and would expect changes in the balance perception. For the passenger in the vehicle, the signals from the vestibular system and the somatic system would be identical to those for the driver. However, the signals from the visual system would have no motion but would show the pages of the reading material. In addition, the passenger would be unaware of the driver turning the steering wheel and so would have no 'control' input (Figure 6.11).

For the driver in Figure 6.11, the four sensory channels provide a consistent cognitive model of the motion environment. For the passenger in Figure 6.11, the sensory channels do not provide a consistent cognitive model, as two of them are detecting motion and two of them are not. It is rare for drivers of cars to get sick, but it is common for passengers in cars or buses to report that reading will make them feel nauseous (Probst et al., 1982).

It is difficult to reconcile motion sickness symptoms with any type of evolutionary advantage, as it does not enhance human capability in any way. It is generally considered that the phenomenon of motion sickness is a side effect of the response of the body to ingesting toxins. Many potentially harmful, even lethal, substances can affect the sensory system before causing permanent tissue damage. For example, alcohol is a toxin that can have fatal consequences if consumed in large enough quantities. Before permanent damage can occur, the effect of the alcohol on the balance system can produce sensory conflicts, whereby the sensation of the 'room spinning' can occur. In this case, it is appropriate for the brain to interpret the combination of signals as injestion of a potentially harmful substance and therefore induce emesis. An alternative sensory scenario would be for an individual to be sat in a chair that is genuinely rotating (e.g., some fairground rides) which could also give a sensation of the 'room spinning'. In this case motion sickness could occur, which would be a misinterpretation of the sensory signals by the brain. Taking an evolutionary timescale, it is more likely to have injested a toxic substance than to be travelling as a passenger or sitting on a rotating chair, and therefore motion sickness has some logical framework.

A model of the genesis of motion sickness is shown in Figure 6.12 (Mansfield, 2005). The inputs from the sensory system as previously discussed are compared to the memory of the expected combination of sensory inputs in the 'comparator'.

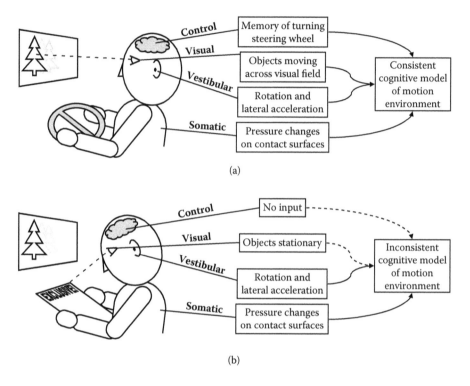

FIGURE 6.11 Illustrative model of the multiple pathways through which a motion environment is perceived: (a) For a driver of a car and (b) For a passenger in the same car reading a newspaper. (From Mansfield, N. J., 2005, *Human Response to Vibration*, Boca Raton, FL: CRC Press.)

If a match is detected then there is no change in the response of the body and homeostasis is maintained. However, if mismatch is detected, then this can trigger motion sickness symptoms, but it also updates the memory. Over time, the memory will become updated such that eventually, when the brain receives such a combination of sensory signals, these are matched with an expected combination of sensory inputs. Once this occurs, the trigger for motion sickness is removed, and the response to the previously nauseogenic stimuli is homeostasis. This habituation can take several weeks of repeated exposures to nauseogenic stimuli to occur.

In studies of the response of the body to motion in the laboratory and in the field, low-frequency motion had been shown to be more nauseogenic than higher-frequency motion. The most nauseogenic frequency of motion is 0.2 Hz (Lawther and Griffin, 1987). For road vehicles, the degree of such low frequency motion is primarily related to the layout of the road and behaviour of the driver. Minimising the amount of low-frequency acceleration will reduce the occurrence of motion sickness. This can be done by selecting routes that maximize straight roads rather than winding roads, and by ensuring that the driver takes corners at a relatively

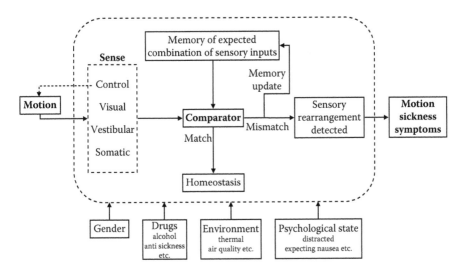

FIGURE 6.12 A model for the genesis of motion sickness that complies with sensory rearrangement theory and illustrates the mechanism of habituation. (From Mansfield, N. J., 2005, *Human Response to Vibration*, Boca Raton, FL: CRC Press.)

slow speed, and does not aggressively accelerate or brake (Turner and Griffin, 1999a, 1999b).

6.7 SUMMARY

For cars, vibration is unlikely to constitute a health risk according to health risk models, if the vehicle is driven on well-maintained roads. Off-road vehicles and heavy goods or public service vehicles can expose drivers to significant vibration. Vibration is one of many factors that contribute to an overall sensation of comfort in a vehicle, and should be considered alongside the posture, seat design and duration of sitting. Vibration accelerates discomfort onset. Vibration control can be achieved through improvements in the design of vehicle seats, through optimisation of the seat foam or, where appropriate, optimisation of seat suspension components. Driver behaviour can affect the vibration to which vehicle occupants are exposed where reduced speeds and careful route selection can minimise vibration exposure.

Workload increases with increases in vibration magnitude even if there is no observable degradation of performance. Primary control performance is not usually affected by vibration for road vehicles under normal circumstances. The operation of secondary controls can be affected by vibration and care needs to be taken to ensure that they can be operated as per their design requirement.

Motion sickness can occur when there is a mismatch between the expected combination of sensory signals and the experienced combination of sensory signals. This is known as sensory rearrangement. Drivers rarely experience motion sickness, but passengers often report motion sickness symptoms. Repeated exposures to a nauseogenic environment usually results in habituation and desensitisation.

REFERENCES

Baker, W. and Mansfield, N. J. 2009. The effects of horizontal whole-body vibration on continuous manual control performance in standing persons. *Proceedings of the 44th UK Conference on Human Response to Vibration*, Loughborough, UK.

Baker, W. D. R. and Mansfield, N.J. 2010. Effects of horizontal whole-body vibration and standing posture on activity interference. *Ergonomics* 53(3), 365–374.

Ebe, K. and Griffin, M. J. 2000a. Qualitative models of seat discomfort including static and dynamic factors. *Ergonomics*, 43(6), 771–790.

———. 2000b. Quantitative prediction of overall seat discomfort. *Ergonomics*, 43(6), 791–806.

———. 2001. Factors affecting static seat cushion comfort. *Ergonomics*, 44(10), 901–921.

European Commission. 2002. Directive 2002/44/EC of the European Parliament and of the Council of June 25, 2002 on the minimum health and safety requirements regarding the exposure of workers to the risks arising from physical agents (vibration), (Sixteenth individual directive within the meaning of Article 16(1) of Directive 89/391/EEC). *Official Journal of the European Communities*, L177.

Griefahn, B. and Bröde, P. 1999. The significance of lateral whole-body vibrations related to separately and simultaneously applied vertical motions. A validation study of ISO 2631. *Applied Ergonomics*, 30(6), 505–513.

Griffin, M. J. 1978. The evaluation of vehicle vibration and seats. *Applied Ergonomics* 9(1), 15–21.

———. 1990. *Handbook of Human Vibration*. London: Academic Press.

Hart, S. G. and Staveland, L. E. 1988. Development of NASA-TLX (Task Load Index): Results of Empirical and Theoretical Research. *Advances in Psychology*, 52, 139–183.

International Organization for Standardization. 1992. Mechanical Vibration—Laboratory method for evaluating vehicle seat vibration—Part 1: Basic requirements. ISO 10326. Geneva: International Organization for Standardization.

———. 1997. Mechanical vibration and shock—Evaluation of human exposure to whole-body vibration—Part 1: General requirements. ISO 2631-1. Geneva: International Organization for Standardization.

Lawther, A. and Griffin, M. J. 1987. Prediction of the incidence of motion sickness from the magnitude, frequency and duration of vertical oscillation. *Journal of the Acoustical Society of America*, 82(3), 957–966.

Mansfield, N. J. 2005. *Human Response to Vibration*. Boca Raton, FL: CRC Press.

———. 2010. Bridging the gap between laboratory trials and the human in context. In Mascie-Taylor, Yasukouchi, Ulijaszek, (eds.) *Human Variation: From the Laboratory to the Field*. Boca Raton, FL: CRC Press.

Mansfield, N. J. and Griffin, M. J. 2000. Difference thresholds for automobile seat vibration. *Applied Ergonomics* 31(3) 255–261.

Mansfield, N. J. and Maeda, S. 2011. Subjective ratings of whole-body vibration for single- and multi-axis motion. *Journal of the Acoustical Society of America* 130, 3723–3728.

Mansfield, N. J., Ashley, J. and Rimell, A. N. 2007. Changes in subjective ratings of impulsive steering wheel vibration due to changes in noise level: A cross-modal interaction. *International Journal of Vehicle Noise and Vibration* 3(2), 185–196.

Marjanen, Y. 2010. Validation and improvement of the ISO 2631-1 (1997). Standard method for evaluating discomfort from whole-body vibration in a multi-axis environment. PhD thesis, UK: Loughborough University.

Marjanen, Y. and Mansfield, N. J. 2010. Relative contribution of translational and rotational motion to discomfort. *Industrial Health* 48, 519–529.

Moseley, M. J. and Griffin M. J. 1986. Effects of display vibration and whole-body vibration on visual performance. *Ergonomics* 29(8), 977–983.

Newell, G. S. and Mansfield, N. J. 2008. Evaluation of reaction time performance and subjective workload during whole-body vibration exposure while seated in upright and twisted postures with and without armrests. *International Journal of Industrial Ergonomics* 38(5–6), 499–508.

Paddan, G. S. 2004. Repeated measures of vehicle vibration during a normal work journey. *Proceedings of 39th UK Group Conference on Human Response to Vibration*, Ludlow, 15th to 17th September.

Paddan, G. S. and Griffin, M. J. 2002a. Evaluation of whole-body vibration in vehicles. *Journal of Sound and Vibration*, 253(1), 195–213.

———. 2002b. Effect of seating on exposures to whole-body vibration in vehicles. *Journal of Sound and Vibration*, 253(1), 215–241.

Probst, T., Krafczyk, S., Buchele, W. and Brandt, T. 1982. Visuelle pravention der bewegungskrankheit im auto (visual prevention from motion sickness in cars). *Archiv fur Psychiatrie und Nervenkrankheiten*, 231(5), 409–421.

Qiu, Y. and Griffin M. J. 2003. Transmission of fore–aft vibration to a car seat using field tests and laboratory simulation. *Journal of Sound and Vibration*, 264(1), 135–155.

Turner, M. and Griffin, M. J. 1999a. Motion sickness in public road transport: Passenger behaviour and susceptibility. *Ergonomics*, 42(3), 444–461.

———. 1999b. Motion sickness in public road transport: The effect of driver, route and vehicle. *Ergonomics*, 42(12), 1646–1664.

7 Thermal Environments and Vehicles

Simon Hodder
Loughborough University, UK

CONTENTS

7.1 INTRODUCTION

With the strong drive across the transport industry to improve passenger comfort, there is interest in optimising the thermal environment. The occupant space of a vehicle is a complex thermal microclimate, which can be prone to large temporal and spatial variation. The space has to be able to provide protection from extreme external environments over a wide range of temperatures, very cold winters (sub-0°C temperatures), and increasingly hot summers (plus 30°C). To be able to design occupant spaces and the control systems for these spaces, which can provide thermally comfortable environments for the passengers, it is important to understand the thermal responses of the human body. This chapter considers the elements that contribute to the occupant response and perception of the thermal environment, and measurement and evaluation of vehicle spaces to ensure the optimum conditions for driver and passengers alike.

Human thermal response to the environment and by extension our perception of thermal sensation and comfort is predominately influenced by six basic parameters. There are four environmental and two personal parameters which directly affect the physiological and human response to the thermal environment and the perception of thermal sensation and comfort.

7.1.1 ENVIRONMENTAL PARAMETERS

The environmental parameters are, air temperature (t_a), mean radiant temperature (t_r), air velocity (v_a) and relative humidity (φ). An understanding of the interaction of these parameters is important to enable designers to produce occupant spaces capable of producing and maintaining thermal comfort.

Air temperature is most often defined as 'the temperature of the air around a person', ISO 7726 (2001). In terms of people, this is the temperature of the air that surrounds them and which is representative of their surroundings. Clothing acts as a boundary between the person and the actual temperature of the air on the other side of the material. This means that the temperature of the air next to the skin is usually different from that of the air surrounding the person.

Mean radiant temperature (t_r) has a strong effect in the occupant space, often greater than would be experienced in the built environment (Madson, Olesen and Reid, 1986). Heat is exchanged between bodies by radiation as well as convection. There is a flow of energy from the hot body to a cooler body. Mean radiant temperature is defined as 'the uniform temperature of an imaginary enclosure in which radiant heat transfer from the human body is equal to the radiant heat transfer in the actual non-uniform enclosure', ISO 7726 (2001). The greatest source of the radiant energy in the human thermal environment is the sun and the specific effects of solar radiation will be considered later in the chapter.

Air velocity is the movement of air across or against a body. This movement is not constant in time, direction or space. For practical purposes, the 'mean' air velocity is often used to define air movement. In confined spaces with directional air flow, it is also useful to consider the turbulent intensity of the air movement.

Absolute humidity of the air, describes any quantity related to the actual amount of water vapour contained in the air. We more commonly refer to relative humidity (φ); this is the actual amount of water vapour in the air in relation to the maximum amount that it can contain at a given temperature. This is often expressed as a percentage. Relative humidity (RH) can be expressed as the ratio of partial vapour pressure of water vapour to the saturated vapour pressure.

It is the interaction of these four parameters that define the thermal environment. Variation in one of the parameters can have a large effect on the thermal response of a person in certain conditions whilst limited effects in others. A high relative humidity (80%) at low air temperature (10°C) has limited effect upon evaporative cooling but the same relative humidity at an air temperature of 30°C will significantly impact a person's ability to lose heat via evaporation.

7.1.2 PERSONAL PARAMETERS

The two personal parameters which influence our response to the thermal environment are metabolic heat production and clothing. The human body takes energy in the form of food which is combined with oxygen, in a process called metabolism. This generates the required energy for the contraction of the muscles during work, blood circulation, breathing and for building body tissues. The metabolic rate increases when tasks are performed to provide the energy required for the work. The body's efficiency is low, the amount of energy that it produces is much greater than the external work requires. This extra energy is transformed to heat. Metabolic rate can be directly measured or more commonly for this type of application estimated (ISO 8998, 2004).

Metabolic rate depends upon the person's age, sex, and body dimensions, (size and weight). To compensate for variations in the body dimensions, the energy produced is often expressed as a function of body area, Wm^{-2}. There can be individual variation in metabolic rates and these may provide significant discrepancies between published mean metabolic rates and actual rates (McIntyre, 1980). In thermal comfort studies, it is common to use established tables (ISO 8996, 2004); these data have been extensively researched and present well-determined estimations of metabolic rate for given tasks. Generally, there is wide agreement that the determined metabolic rates for sedentary and light work, (1 to 2 Met), are accurate for the study of thermal comfort (ASHRAE, 1993), and this makes them appropriate for evaluation of vehicle environments.

Traditionally, metabolic heat production is measured in Met (1 Met = 58.15 Wm^{-2} of body surface). The estimations assume an average male adult, (70kg, 1.75 m stature) has an approximate surface area of 1.8 m^2, and a person in thermal comfort with an activity level of 1 Met will thus have a heat loss of approximately 100 W. Driving under normal road conditions would have a Met rate of 1.2 (70 Wm^{-2}). Driving over rougher terrain may give a slightly high metabolic rate.

Clothing can have a large effect on the ability of a person to achieve or maintain thermal comfort. Clothing provides a thermal resistance between the body and the environment. For most of us, clothing's primary function is to help maintain the body in an acceptable thermal state. Clothing is often adapted to the environment that we live in at a particular time. Clothing reduces the body's heat loss and is therefore, classified according to its insulation value. Clothing's insulation unit is the Clo, where 1 Clo would be a typical business suit with shirt. The technical units (m^2°C/W) of clothing insulation is also seen, where 1 Clo = 0.155 m^2°C/W.

TABLE 7.1
Summary of Clothing Insulation Values

Clothing Item	I_{clu}	Clo Value
T-shirt	0.09	0.014
Short sleeve shirt	0.09	0.029
Normal, long sleeves	0.25	0.039
Flannel shirt, long sleeves	0.3	0.047
Light-weight trousers	0.2	0.031
Overalls	0.28	0.043
Thin sweater	0.2	0.031
Jacket	0.35	0.054
Fabric-covered, cushioned, swivel chair	0.1	0.016

The total Clo value can be readily calculated if we have knowledge of the individual components of a person's ensemble, a person's dress and the Clo values for the individual garments are known, by simply adding the Clo values together. Table 7.1 contains a list of selected clothing items and their corresponding Clo values, (ISO 9920, 2007).

When calculating Clo values, it is important to remember that upholstered car seats can have a considerable effect, and reduce the heat loss from the body. They act as an additionally layer of less permeable clothing and must therefore be included in the overall calculation. Dependent upon the coverage of the seat, the additional insulation value offered can be as high as 0.2 Clo.

It is the interaction of these six parameters that determine a person's response and perception of their thermal environment. These environmental factors can act upon the person as they strive to maintain a steady deep body temperature. Humans are homotherms and it is essential for them to maintain an internal body temperature around 37°C. Even small deviations, around this temperature can cause physical stress. The body uses internal heat generation and heat exchange with the surrounding environment to regulate the internal temperature. This heat balance is often referred to as steady state (Fanger, 1970; Parsons, 2003). Steady-state heat balance implies that a constant temperature is maintained, however humans are in a continual energy flux between themselves and their environment. As the body temperature is kept in a constant range rather than a specific single temperature, then the term dynamic balance is more appropriate, (Parsons, 2003).

If the heat energy entering the body is greater than that leaving the body, the temperature of the body rises. This results in vasodilatation and sweating as the core temperature is elevated. If the heat leaving the body is greater than that going into it, the temperature falls. This results in vasoconstriction and shivering, driven initially by the decrease in skin temperature followed by a decline in core temperature. Getting the balance right can optimise the thermal comfort of the vehicle occupant.

7.1.3 THERMAL COMFORT

Thermal comfort is defined in ISO 7730 (2005) as being 'That condition of mind which expresses satisfaction with the thermal environment'. This is a definition on which there is common agreement (ASHRAE, 1993), it is also a definition which is not necessarily easily translated into physical parameters. Thermal comfort is therefore a subjective response, which is derived from the effect of the physical environment on the physiological responses of the body. It is the interaction of environmental and personal parameters that makes the evaluation and prediction of thermal comfort difficult. Thermal comfort can be determined through the use of comfort indices.

7.1.4 THERMAL COMFORT INDICES

A comfort index is a simple number that can be used to describe the thermal environment and its effect on a person. Over the last century numerous thermal comfort indices have been proposed and fall into three main categories; direct, empirical and rational indices.

Direct indices use environmental measurements with simple instruments that respond to the thermal environment in a similar way to humans. The environmental measure that the majority of people readily understand most easily is air temperature, however, it is a poor indicator of the thermal environment. If we think of the weather report on a summer's day, the air temperature is given on its own. Air temperature alone does not give the full picture of how the environment will be perceived; it does not consider the additional effect of solar radiation or air movement which can have a significant effect. For this reason most direct indices incorporate at least two environmental measures.

One of the most widely used direct indices is the Wet Bulb Globe Temperature (WBGT) Index (ISO 7243, 1989). It is primarily used as an indicator of heat stress conditions and is widely used around the world in industrial environments. It is an approximation of corrected effective temperature (CET) with a correction for solar radiation. The WBGT contains no measurement of air velocity, although increased air movement in solar conditions reduces globe temperature and the natural wet bulb temperature, so there is some correction for the effect of cooling by air flow, (Kerslake, 1972). This has potential for assessment of vehicles in extreme hot environments.

Direct indices can provide a good, quick approximation of the thermal environment. These indices should be used with care, as they have limitations if applied over a wide range of environmental conditions.

Empirical indices are developed from data collected from human subjects in known environments. Through such experiments over a wide range of conditions, it is possible to determine how people will feel in them.

7.1.4.1   Effective Temperature

This is an early index, it is not a temperature as such, but an 'arbitrary index of the sensation of warmth experienced as a result of air temperature, humidity and air motion. It combines these three factors into a single value' (Yaglou, 1927). Its underlying principle is that changes in any of the three factors may vary greatly as long as the combined effect remains the same. Hence, an increase in air temperature must be compensated for with a corresponding decrease in relative humidity or increase in air velocity. Effective temperature does not take into account radiation, although a correction can be made by using a 150 mm black globe thermometer (CET). Such indices require the user to be experienced with the principles behind them. Often the measured temperature will have to be interpreted using psychrometric nomograms.

7.1.4.2   Equivalent Temperature (T_{eq})

There have been a number of different versions of equivalent temperatures presented. Bedford (1936) analysed a number of indices, including an earlier version of T_{eq} derived by Dufton in 1932, against sensation votes and settled up a 150 mm black globe thermometer to represent a person. The equation he derived for calculating T_{eq} was:

$$T_{eq\,(Bedford)} = 0.522t_a + 0.478t_r - 0.21\sqrt{v}(37.8\text{-}t_a)$$

In a modern context, equivalent temperature (t_{eq}) is defined as temperature of a homogenous space, with mean radiant temperature equal to air temperature and zero air velocity, in which a person exchanges the same heat loss by convection and radiation as in the actual conditions under assessment, (ISO 14505-2, 2006). This equivalent temperature is derived from measurements taken made with thermal manikins (see Section 3.2).

> **Rational indices** are based upon the principle of heat balance. If a body is to remain at thermal equilibrium, then the heat energy into the body must be balanced with the heat energy leaving the body.

$$M \pm K \pm C \pm R - E = S$$

where
  M = metabolic rate (Wm^{-2})
  W = external work
  E = evaporation
  R = radiation
  C = convection
  K = conduction
  S = heat storage

If the body is in heat balance, then $S = 0$. Therefore, any rational index will need information about the environment and activity of the people in it, in terms of the six parameters, (t_r, t_a, v, φ, Met and Clo).

There are a number of rational indices that have been developed over the last 70 years, as scientists and engineers have tried to quantify an environment in terms of a single number.

7.1.4.3   Operative Temperature (T_o)

Gagge (1940) defined operative temperature as the temperature of a uniform black enclosure in which a human occupant would exchange the same amount of heat by radiation and convection as in the actual non-uniform environment, and is defined by the equation:

$$T_o = T_a + h_r (T_r - T_a)/h_{cr}$$

where
  h_r = linear radiation exchange coefficient.
  H_{cr} = is the combined coefficient, ($h_r + h_c$), with h_c being the convective heat transfer coefficient.

Operative temperature is effectively a weighted average of mean radiant temperature and air temperature. It works well when mean radiant temperature does not deviate significantly from air temperature. This is why it is still commonly used within buildings as a measurement. However, if the constituent components of t_r vary greatly, then it becomes necessary to consider the relative absorptance of the body surfaces with reference to the dominant radiative heat source.

7.1.4.4   Thermal Comfort and the Comfort Equation

For us to understand the requirements for the occupant environment it is important to understand how the occupants perceive thermal sensations, determine their thermal comfort and ultimately, how satisfaction with the environment can be achieved.

The variety of indices, which had been developed over time proved problematic and difficult to use in practical situations by people who were not experienced in their application. This led to Ole Fanger (1970) devising a simple thermal comfort index, the predicted mean vote (PMV) model. It was envisaged that the index should incorporate the six basic parameters into a single equation, accounting for the interactions between each of the variables. The major shift from previous philosophies regarding thermal indices was to define comfort in physiological terms of the person, rather than that of the environment that the person was in. The rationale that a person senses changes in skin temperature rather than in air temperature meant that the person experiencing the conditions was the important factor, and what would now be considered a user-centred approach.

Fanger set out 3 conditions for thermal comfort:

The body must be in thermal equilibrium (heat balance).
Mean skin temperature is at a level appropriate for thermal comfort.
Sweating is at a preferred rate for comfort.

If we consider these requirements in greater detail, it is possible to understand the philosophy behind this model further. When the body is in thermal equilibrium, the heat losses and heat gains to the body are the same, giving zero heat storage. This implies the 'steady state' thermal condition, if the body is not in heat balance, then it will quickly adapt to maintain this balance. If it is cooler than required shivering and vasoconstriction will start to increase metabolic heat production and reduced heat losses from the body surface. If it is too hot, sweating and vasodilatation occur to increase heat losses from the body. It is not sufficient for comfort to merely maintain heat balance, as this can be done in conditions which are far from thermally comfortable (McIntrye, 1980).

Thermal neutrality is achieved when the environmental parameters placed into a thermal index give a thermal sensation response of zero (neutral). It is therefore possible to determine combinations of air temperature, mean radiant temperature, air velocity and relative humidity that the majority of occupants would have a thermal sensation response of neutral. In these conditions, the occupants should predominately feel, neither, cool nor warm. Thermal neutrality differs from thermal comfort in that people can be comfortable away from temperatures that would not be considered thermally neutral.

Thermal sensation is related to mean skin temperature. The skin is the thermal interface of the body; temperature receptors are widely distributed over the whole body. Some areas of the body have higher concentrations of temperature receptors than others, these are predominately the fingers and toes, and to a lesser extent the hands and feet (Clark and Edholm, 1985). The receptors respond to changes in temperature, which are transmitted via the afferent nerve to the hypothalamus, the body's 'thermostat'. So, it can be seen that these sensations have a direct response on the thermoregulatory control mechanism in the brain. Gagge, Stolwijk and Hardy (1967) presented the relationships between mean skin temperature, thermal sensation and comfort. They found that a mean skin temperature of approximately 33°C will provide neutral thermal sensation with subjects being comfortable. Deviations from this 'comfort' skin temperature have a rapid effect on thermal sensation, but decreases in comfort do not occur so rapidly.

As the metabolic rate increases, the mean skin temperature will decrease; Fanger determined mean skin temperature with the following formula;

$$t_{sk} = 35.7 - 0.032 \, H/A_{du}$$

where
   H = internal heat production of the body (Wm^{-2})
   A_{du} = Dubois body surface area

Using this, it is possible to see the effect of an increase in metabolic activity on mean skin temperature (t_{sk}). Where a metabolic rate of $50 Wm^{-2} = t_{sk}$ 34°C, whilst, $150 \, Wm^{-2} = t_{sk}$ 31°C.

Sweat secretion at thermal neutrality was found to be zero. As metabolic activity increases, moderate sweating is necessary to maintain heat balance and thermal comfort. It had been assumed that for comfort conditions, that a mean skin temperature of 33 to 34°C with no sweating was required, this is only the case for sedentary tasks. Taking these criteria into consideration, Fanger proposed a theoretical heat balance equation for comfort, based on the criteria that for constant, moderate thermal conditions, the body's heat production would be equal to its heat dissipation.

This comfort equation only gives information about how the environmental and personal variables can be combined to provide optimal thermal comfort. It is from this point that Fanger derived an index that would provide a prediction of a person's thermal sensation for any given combination of environmental conditions, clothing and activity levels.

7.1.5 Predicted Mean Vote

Considering that the body is capable of maintaining heat balance over a wide range of environmental variables via the use of thermoregulatory control mechanisms (vasodilatation, vasoconstriction, sweating and shivering), thermal comfort occupies only a small part of this range. As the body deviates from comfort conditions, the load on the thermoregulatory mechanism increases. Fanger proposed that the thermal sensation of a person at a known activity level is a function of the thermal load (L). The thermal load can therefore be considered as the physiological strain upon the thermoregulatory mechanisms to maintain comfort.

The following equation was proposed as a description of the relationship between thermal sensation and physiological strain;

$$Y = f(L * H/A_{Du})$$

where
   Y = mean vote on the ASHRAE thermal sensation scale.
   f = function relating mean vote to thermal load and internal heat production.
H/A_{Du} = internal heat production.

The ASHRAE Scale	
7	Hot
6	Warm
5	Slightly warm
4	Neutral
3	Slightly cool
2	Cool
1	Cold

From this theoretical point, it was necessary to establish the nature of the functional connection between L and H/A_{Du} and thermal sensation. To establish the relationship, a database of 1396 sensation votes for a variety of environmental conditions and activity levels, the same clothing level was maintained throughout all experimental conditions. The component Y becomes the predicted mean vote (PMV), a single figure output that can be easily related to how people would perceive a given environment. This comfort index has become the dominant model for assessing and evaluating human thermal comfort. It was adopted as the preferred method for assessing thermal comfort in moderate environments by the International Standards Organisation (ISO 7730, 2005). Its robustness has been shown over years and is used to provide design guidance for a variety of human thermal environments.

7.1.6 ADAPTIVE THERMAL COMFORT

In the last 20 years, the adaptive model of thermal comfort has also become popular. The theory behind this model is that factors beyond the traditional four environmental parameters (t_a, t_r, φ, and v) and the personal parameters (Met and Clo) influence a persons perception of thermal sensation (de Dear and Brager, 2001). It is suggested that exterior temperature amongst other factors, previous life experiences, acceptability and preference, will affect the thermal comfort perceptions of a building's occupants.

It is interesting to note the adaptive philosophy, in certain contexts, a stressful environmental parameter, can become a pleasant one in another context, for example, solar radiation, in the summer, when t_a is high, can add to the heat stress, whilst in the winter, when t_a is low, it can give a pleasant feeling of almost thermal neutrality. Vehicle environments, and particularly, cars offer a high level of adaptive opportunity to the occupant. The occupant has the ability to control the temperature of air entering the space, air velocity by both the ventilation system and opening and closing windows.

7.1.7 SOLAR RADIATION

One of the elements that can substantially affect the radiant environment within a vehicle is the sun. Solar radiation has been shown to have a considerable impact on occupant discomfort, (Rohles and Wallis, 1979; Parsons and Entwhistle, 1986; Hodder and Parsons, 2007).

Solar radiation is electromagnetic radiation emitted from the sun. The sun is a black body emitter with a surface temperature of approximately 5800 to 5900 Kelvin. The electromagnetic spectrum ranges from cosmic, gamma and x-rays, ultraviolet (UV), photosynthetically active radiation (PAR), commonly known as 'visible' radiation, infrared (IR), microwaves, through to very-high-frequency (VHF) radiation, (McKinlay et al., 1988). As the solar radiation passes through the earth's atmosphere; some of it is absorbed, relative to its wavelength. Not all solar radiation passes through the earth's atmosphere. Electromagnetic energy with wavelengths less than 2.8 µm, which includes the lower part of the UVB spectrum, is absorbed by ozone. Clouds also reflect a significant fraction of solar radiation back to outer space, whilst the remainder reaches the surface of the earth in both a direct and diffused form. The intensity of radiation depends upon the thickness of air, which it must penetrate. This is determined by the earth's rotation about its axis and it revolution around the sun. The irradiance of the sun just outside the earth's atmosphere is approximately 1350 Wm^{-2}. The levels detected on the earth's surface are lower than

TABLE 7.2
Definitions of Spectral Radiation Wavelengths

Spectral Type	Wavelength
UVB	280 to 315–320 nm
UVA	315–320 to 380–400 nm
PAR/VISIBLE	380–400 to 760–780 nm
IRA	760–780 to 1400 nm
IRB	1.4–3.0 μm
IRC	3.0 μm–1 mm

this. The earth's atmosphere attenuates the amount of energy actually reaching the surface, the maximum levels measured are in the region of 1000 Wm^{-2}. These high levels of direct solar radiation are rare and occur generally where the sun has a high altitude, and the point is high above sea level.

Factors that need to be considered when predicting the intensity of solar radiation are; solar incidence (azimuth and altitude of the sun), geographical location (altitude and latitude), the cross-sectional area of the body exposed to the sun, cloud cover, dust, carbon dioxide and water vapour and the surrounding terrain.

The solar spectrum that makes its way through the atmosphere is ostensibly divided into three regions, UV, PAR and IR, which are divided into sub-sections (see Table 7.2).

Only a small section of the spectrum is visible to the human eye. But this contains 45% of the energy emitted as well as the peak levels of energy intensity, (Givoni, 1976), with UV accounting for 5% and IR for 50%.

In terms of the effects of radiation types on the human body, UVA and UVB have the most significant effect on people. UVA is the least harmful of the ultraviolet radiation types to the human skin with the main effect being to increase existing tanning. It is the radiation in the UVB band that has most impact upon the skin. Exposure to it rapidly induces tanning, with short exposures tending to burn exposed skin. UVB radiation is in general attenuated by glazing, this means that for the most part it has limited impact on vehicle occupants.

Evaluating the impact of solar radiation on people and particularly vehicle occupants is complex. Air temperature, relative humidity and air velocity for the most part can be accurately and repeatedly replicated in climatic chambers and wind tunnels. Mean radiant temperature and particularly solar radiation are more difficult to reproduce in a controlled environment.

Hodder and Parsons (2007) conducted a series of laboratory studies investigating the impact of solar radiation on vehicle occupants. Using a solar radiation lamp and a climate-controlled chamber, they determined that intensity of the radiation was the largest determinant of thermal discomfort, with a solar radiation load of 200 Wm^{-2} producing a thermal sensation shift of 1 scale point (thermal sensation shifts from neutral to slightly warm) in an otherwise thermal neutral environment. Direct solar radiation of different spectral qualities but with the same level of radiant energy on the person, in this case 400 Wm^{-2} does not produce significantly different thermal sensation responses. They also evaluated four different glazing types; clear, tinted, tinted laminated and IR laminated, with the same solar load on the exterior of the glazing 1000 Wm^{-2} and found that this did produce variations in the thermal sensations reported.

These results show that the type of glazing used in vehicles does make a difference to the thermal sensation and comfort felt by the occupants. But, it is the reduction in the transmission of total radiation through tinted and IR treated glazing rather than spectral content of the radiation that is the important factor.

Clothing is a factor that also needs to be a consideration when assessing the impact of solar radiation. The colour of clothing has been found to impact upon people's responses to solar radiation (Nielsen, 1990), although the transmission and reflective properties of the clothing also have an effect. The thickness of clothing material can help to reduce the transmission of solar energy to the body (Roller and Goldman, 1968). Clark (1981) found that when the body is directly irradiated, hair and clothing temperature increased by 15 to 18°C whilst exposed skin increased by just 5 to 6°C.

Rational models based upon heat balance equations have been dominant over the last 40 years, and will continue to be the preferred method of quantifying a person's thermal state in a given environment. In addition to a simple model like the PMV, there are more complex physiological models which use passive and controlling systems. An example of this would be the 25-node model developed by Stolwijk and Hardy (1977). This included elements representing the body (limbs torso, hand, feet and head), core, muscle, fat and skin layers and additional blood compartment, this is the passive system which models the physical human body and the heat-transfer in it. The control system simulates the thermoregulatory responses of the central nervous system. With the development of computer processing capabilities more complex models of thermal regulation have been developed (Fiala, Lomas and Stohrer, 1999, 2001). This model has many more facets to its design than that of the earlier models; a greater number of body segments, each with more tissue layers, which allows for greater interaction with an environment, particularly thermal asymmetries.

In recent years, the development has been towards coupled models. Placing human thermal models within computational fluid dynamics (CFD) programmes so that the models provide complimentary feedback to each other. An early example of this was a predictive modelling programme called INKA/TILL for the evaluation of vehicle occupant spaces. This software used two programmes: INKA, to calculate environmental parameters; and TILL, to calculate thermal responses of the occupants, (Zimmy et al., 1999). Actual environmental t_{eq} measurements were compared with predicted responses, but with limited success. This was considered to be due to poor basic parameters, suggesting that is important to know in detail the environmental conditions within the occupant space.

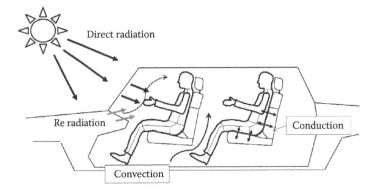

FIGURE 7.1 Avenues of heat exchange within a vehicle.

7.2 VEHICLE ENVIRONMENTS

When a passenger enters the vehicle and takes their position, there are a number of avenues of heat exchange that are established (Figure 7.1):

Convection within the occupant space through air movement, primarily driven by a ventilation system or open windows.

Conduction via the back and thighs in contact with the seat, and feet in the footwell, and for the driver contact with the steering wheel.

Radiation either by sunlight directly onto the occupant, re-radiation from inner surfaces (dashboard) being heated up or cold surfaces (windows).

Depending upon the external environment at the time, these avenues of heat exchange could either contribute to heat gain to the occupant, vehicle is hotter than the person, or heat loss, vehicle is colder than the person. When a vehicle is parked, the occupant space will soon equilibrate with the wider surrounding environment. For environments where the vehicle is exposed to direct solar radiation, the thermal environment in the vehicle will exceed that of the surroundings. It can be common for a car soaked on a summers day (outdoor t_a 34°C) to have air temperatures in the unventilated occupant space be 65°C and have surface temperatures up to 100°C (Atkinson, 1987; Mezrhab and Bouzidi, 2006). Once the doors are opened, the air temperature drops quickly with the hot air moving out, but the heated surfaces will remain hot for some time and continue to radiate heat into the space.

Air velocity is the environmental parameter that the occupants have the most control of via the ventilation system. It can be actively used to improve the thermal comfort depending upon their requirements. Relative humidity is considered to be the parameter that needs the least amount of consideration in such environments (Temming, 1980).

These factors make external air temperature and mean radiant temperature the principle factors in determining the initial thermal environment when the occupant enters the vehicle.

It is possible to mitigate the impacts of these avenues of heat exchange over time by the addition of HVAC systems. But for these systems to produce a thermally comfortable environment, it can take 10 minutes or more. During this time the occupants are likely to feel a level of thermal discomfort and dissatisfaction with their surroundings. These changes from one thermal environment to another are considered as transients.

Traditionally heating systems within vehicles have utilised heat generated from the engine, in conjunction with the ventilation system to provide a method of warming the compartment. The drive to smaller economic engines means that there is often less heat available as a by-product to help control the occupant space.

7.2.1 Transient Thermal Comfort

Traditionally the majority of thermal comfort research and design guidelines has been based around steady-state built environments. A vehicle that has been parked will have an internal thermal environment similar to that of the external environment. For most cases, this will tend to be either much cooler or hotter than the range of acceptable thermal environments, nominally between 18 and 26°C.

Transient thermal comfort relates the sensation experienced by a person moving from one thermal condition to another. The thermal sensation response can change immediately although physiological changes in mean skin temperature and core temperature respond more slowly (Gagge, Stolwijk and Hardy, 1967). Gagge et al. also suggest that for transient responses, air temperature is a strong indicator of how the person will feel.

The time it takes for the occupant space to become a thermally acceptable environment can vary from the occupants entering the vehicle to being thermally acceptable with the external air temperature.

Oi et al. (2012) found that heated seats can improve occupant comfort during warm up periods if the ambient air temperature is below 15°C. Local sensation felt on the feet was also increased 15°C and above, whilst at air temperatures below that, the heated seat mitigated a decline in skin temperature of the toe. The response of the extremities (hands and feet), in cold environments can have a strong influence on the level of discomfort felt, so this could be a useful tool in helping to limit any thermal discomfort until the occupant space is at optimum temperature.

7.2.2 Seating

The interaction with the seat provides a significant avenue for heat gains and losses. Seats can be made of a variety of materials and can be designed to incorporate heating and cooling systems.

Brooks and Parsons (1999) found that it was possible to increase overall thermal sensation and reduce thermal discomfort of occupants at low air temperatures (5°C) when using an electric filament heated seat. This study allowed the participants to self-select a seat temperature to help maintain their thermal comfort. Notably, hands and feet were a dominant source of discomfort.

Madsen (1994) found in a study with a thermal manikin that ventilated seats could improve the removal of heat. Heat losses from the body parts in contact with the seat surface were greater than those for standard un-ventilated seats. In a larger study, Zhang et al. (2007) looked at the effect of both seat heating and cooling on thermal sensation in summer and winter conditions. The seat temperature was maintained with a controlled water pump system. They found that with temperature controlled seats that steady state thermal sensation was reached in 11 minutes. This is a much quicker response that that normally witnessed in air controlled environments. The use of conduction heating/cooling systems can enable occupants to reach a thermally acceptable state is a much shorter time. In extreme cold environments, this method of heating could have substantial benefits in terms of retarding the onset of whole body thermal discomfort.

The localised heating can be used to reduce thermal discomfort at lower cabin space temperatures. They can also offer much quicker warm-up times that the heating, ventilation, and air conditioning (HVAC) system of a car. Personalised control of the environment has become a popular concept in recent years. Rather than design the occupant space to be uniform and to elicit a mean thermal sensation for a population, design the individual spaces to be controlled by the individual.

Fung and Parsons (1993) undertook an extensive study of different seat materials. Subjects were exposed to a warm environment (34°C t_a, RH 35%), and rated their thermal sensations. Hydrophillic seat coverings were found to be the most satisfactory. Hydrophillic materials aid in the transport of moisture away from the surface of the seat, allowing the sweat to be pulled away from the skin and surface of the clothes. The seats that performed worst were ones with impermeable barriers either in the seat covering material or in the foam used for upholstery. Fung (1997) later analysed the subjective data in conjunction with further laboratory-based experiments on seating materials. It was possible to rank seating materials with reference to their various properties for removing moisture, it was not possible to produce a specification for the most suitable materials that provide good thermal comfort. The widely varying conditions within the occupant space, various clothing ensembles worn, duration of time seated and inter-personal preferences made it very difficult to deduce the ideal seat materials. This could be seen in a study by Cengiz and Babalik (2007), who evaluated the thermal comfort of three different vehicle seats in a field trial. They found no significant difference between the seats under real world conditions.

7.3   VEHICLE ASSESSMENT

New vehicle occupant spaces often need to be evaluated to assess the effectiveness of climate control systems. Two main methods are now available, assessment via thermal manikin and via human subject evaluation.

For either method, it is important to characterise the vehicle space. For this, measures of air temperature, mean radiant temperature, relative humidity and air velocity are needed as a minimum. ISO 7726 (2001) offers guidance for the selection of measurement devices and specifies techniques and operating guidelines.

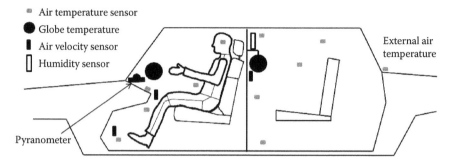

FIGURE 7.2   Distribution of environmental measurement sensors for the evaluation of a vehicle occupant space.

Air temperature can be measured in a number of ways. Traditionally, mercury in glass thermometers has been used to measure air temperature, but these are impractical in modern dynamic situations. For the quantification of a vehicle space, it is important to be able to measure in multiple locations and record data over time. The primary sensors which are used now are thermocouples, thermistors or resistance thermometers/platinum resistance thermometers (PRTs). These can be discreetly placed in the occupant space and record to a data acquisition system. Thermocouples have very rapid response times but do not have the accuracy of the thermistors or PRTs.

Ideally, the sensor heads should be shaded to prevent any direct heating of the sensor by radiation. For each occupant seat, a temperature profile should be established. Vertical temperature gradients between the feet and head can lead to discomfort. A gradient of 4°C between the footwell and the vehicle roof would result in potential 10% dissatisfaction (ISO 7730, 2005).

One of the issues with quantifying the thermal environment of the car is the limited space available for placement of measurement equipment. Ideally, the measurements would be made in the space where the driver or passenger would sit. This is certainly not practical when doing evaluations with either a thermal manikin or human subjects, so the sensors need to be close but not interfere with the subject. Principally, a vertical array of air temperature sensors should be used, representing the footwell, knee/thigh area, torso and head, (see Figure 7.2). It is also important to measure the inlet temperature of the air leaving the ventilation system; this will require a number of sensors to accurately capture the profile around the vehicle.

Mean radiant temperature is commonly quantified with a 150 mm diameter black globe thermometer. The globe temperature value when combined with simultaneous measured values for the air temperature and velocity surrounding the globe enables an approximate value of the mean radiant temperature to be derived. Globe thermometers are available in smaller diameters, and corrections for this can be made in the equations. The placement of globe thermometers within an occupant can be problematic due to the limited space available. Ideally, several globe thermometers should be used. In addition, a pyranometer should also be included to measure direct radiation entering the occupant space. A pyranometer is a radiometer which measures solar radiation approximately from 300 to 2,800 nm, which includes the visible and near infrared spectrum and gives a value of the energy in Wm^{-2}.

Air velocity is a quantity defined by its magnitude and direction. Air velocity is a problematic environmental parameter to measure due to the vectorial characteristics and rapid and strong temporal fluctuations (Gameiro da Silva, 2002). Therefore, it is necessary to record the fluctuations; this will enable a mean air velocity to be determined. The standard deviation of the mean air velocity can be used to determine the turbulence intensity; this gives a measurement of the fluctuation (ISO 8998, 2004).

Air velocity is measured with an anemometer. There are a number of different types; directional anemometers, vane, cup, hot wire and omnidirectional anemometers; hot sphere. Air velocity should be measured close to the person in the proximity of where the ventilation system introduces heated/cooled air into the occupant space as well as in a central position.

Relative humidity is measured with a hygrometer. Traditional, hygrometers used the relationship between dry and aspirated wet bulb temperature and psychometric charts for the determination of relative humidity. Dew point and lithium chloride hygrometers, which record directly to data acquisition systems are more commonly used now. Taking a measurement in the centre of the space will suffice for most evaluations, although it may sometimes be prudent to also measure the relative humidity of the air entering the space via the heating/cooling vents.

As well as the four basic parameters, it is also useful to measure surface temperatures within the vehicle, particularly dashboard and internal roof temperatures as well as the shaded external air temperature.

7.3.1 CHAMBER STUDIES VERSUS FIELD EVALUATION

A substantial body of the work investigating occupant thermal comfort, the factors influencing it, and the HVAC systems has been undertaken in laboratories and climate chambers. This enables the researcher to great control over the individual variables, e.g., air temperature, humidity and so forth, but these test environments lack the authenticity of the real world. Moving from the laboratory/chamber to the field will increase the face validity of the study being conducted. But this will come at the cost of a reduced level of control over the studied environmental conditions. Cengiz and Babalik (2007) found no difference between seat materials in a real world evaluation. This could in part be due to the variation in environmental parameters introduced by undertaking the studies over a number of different days. It may also suggest that some of the thermal benefits of materials seen in laboratory studies may be limited when placed into the far more complex real world scenarios.

7.3.2 THERMAL MANIKINS

Thermal manikins have been used as evaluation tools for vehicles for over 25 years, (Wyon et al., 1985; Madsen et al., 1992; Matsunaga et al., 1993; Nilsson et al., 1997; Holmer, 2004). Thermal manikins are human body forms which

FIGURE 7.3   Thermal manikin 'Victoria' in a car.

are heated where the surface can be heated so as to simulate the heat transfer between humans and their thermal environment.

Typically, a manikin will consist of a number of heated body zones, in which the temperature can be controlled and monitored (Figure 7.3). The number of body zones can vary, the more individually heated and controlled zones the manikin has, the greater the information about heat losses (radiative, convective and conductive), can be gathered.

Manikins have been used to evaluate vehicle environments with a modern variation of the equivalent temperature (t_{eq}). This is described in *Evaluation of Thermal Environments in Vehicles—Part 2: Determination of Equivalent Temperature* (ISO 14505-2, 2006). The equivalent temperature is a physical quantity that integrates the independent effects of convection and radiation on human heat exchange. This relationship is best described for the overall (whole body) heat exchange. The standardised definition of t_{eq} applies only for the whole body. Therefore, the definition has to be modified for the purposes of ISO 14505-2: t_{eq} does not take into account human perception and sensation or other the subjective aspects. However, empirical studies show that t_{eq} values are well related to the subjective perception of the thermal effect. Nilsson et al. (1997) did a comparison of two, 33-zone manikins in both summer and winter conditions in a vehicle cabin in a climatic chamber. Comparing the manikin results with actual subjective human responses to the same environment, they found that, for local body parts, results from the manikins correlated well with subjective responses.

Nilsson et al. (1997) in their comparison of two manikins found that there were differences between their outputs. This was as a result of the variation in the manikins shape/form and their control systems. This means that direct comparison of data from manikins of different build is not possible and may account for variation in results. Another factor that also needs to be considered is that when using a thermal manikin to evaluate a vehicle environment, it has no active cooling mechanism like a human. If the environment is at a greater temperature than that of the manikin surface, typically 34°C, then it will not be able to provide heat loss data. This would also be the case if there was local direct heating by solar radiation, for example in field evaluations in hot environments, the thighs might be expected to heat up more with direct irradiation and this might lead to erroneous data.

7.3.3 Human Subject Trials

Mathematical and physical models and the thermal indices can provide methods of developing and testing environments. Their validity is based upon having good actual data from the type of environments that they hope to model. These databases can often be limited in size for specific environmental conditions and application, that is, cars, trains, buses, aircraft. The number of field studies with human subject data reported is limited, although the amount of confidential commercial data may be much larger. With new vehicles and climate control systems, their effectiveness in achieving optimal thermal conditions can be best evaluated with user trials. Often the best way to determine if people are satisfied with their environment is to ask them. People are powerful assessment tools and can give a great deal of feedback on a product, system or environment.

User evaluations have been around for a long time, with researchers often developing their own techniques, protocol and subjective scales. In recent years, there has been a drive by ergonomists to help standardise measurement and evaluation methods for research. This enables researchers and practicing ergonomists to use scientifically valid protocols and measurement scales for evaluation of environments. ISO 14505-3 *Evaluation of the Thermal Environment in Vehicles, Part 3: Evaluation of Thermal Comfort Using Human Subjects* (2006) is a standard focused directly at the subjective evaluation of cars.

The standard provides general principles for the evaluation of a vehicle, giving information about selection of subjects, subjective scales and protocol guidance.

The selection of subjects should represent the expected user population; age, gender driving experience and anthropometry should all influence the selection. The number of subjects should be sufficient to provide scope for statistical analysis. Methodological advice on experimental protocols and selection of test conditions is given.

Subjective scales are given, the ISO thermal sensation scale; this indicates how the subject feels now. For extreme environments, it may be appropriate to extend the scale from 7 to 11 points, to gain more detail about the environment.

ISO Thermal Sensation Scale		Extended Thermal Sensation Scale	
		+5	Extremely hot
		+4	Very hot
+3	Hot	+3	Hot
+2	Warm	+2	Warm
+1	Slightly warm	+1	Slightly warm
0	Neutral	0	Neutral
−1	Slightly cool	−1	Slightly cool
−2	Cool	−2	Cool
−3	Cold	−3	Cold
		−4	Very cold
		−5	Extremely cold

Additional subjective scales, uncomfortable, stickiness and preference, are also included.

'Uncomfortable' Scale		Stickiness Scale		Preference Scale	
3	Very uncomfortable	3	Very sticky	+3	Much warmer
2	Uncomfortable	2	Sticky	+2	Warmer
1	Slightly uncomfortable	1	Slightly sticky	+1	Slightly warmer
0	Not uncomfortable	0	Not sticky	0	No change
				−1	Slightly cooler
				−2	Cooler
				−3	Much cooler

The uncomfortable and stickiness scales focus on the negative aspects, associated with thermal discomfort and sweating, respectively. The preference scale gives an indication of how the person would like to feel. This is important, as they may report being 'warm' but that they require 'no change', indicating that they are satisfied with the environment.

Overall measures of satisfaction and acceptability, using forced 'yes' or 'no' answers will enable overall percentages to be derived.

In addition, objective measures can also be taken, in particular skin temperature. Mean skin temperature (T_{sk}) is also an important factor in determining heat loss. During evaporative thermoregulation, skin temperature changes slowly with the ambient temperature. When environments are cool, skin temperature reacts strongly as a function of the ambient temperature, making it a good predictor of thermal sensation (McIntyre, 1980). In hot environments, skin wettedness is a good predictor of discomfort during regulatory sweating. Skin wettedness is the ratio of the actual evaporative loss at the skin surface to the maximum loss that could occur in the same environment. It does not imply anything about the rate of evaporative loss, but relates to the perception of sweating and discomfort in the heat and links directly to the subjective stickiness sensation.

Trials should be set up with a specific aim, that is, evaluating glazing or an air conditioning system, and the experimental protocol designed to focus on this aim. In field trials, it is often difficult to replicate the external environmental conditions on a day-to-day basis, although in some parts of the world it is possible to have consistent weather conditions over a period of weeks. Climate control settings should be noted and fixed and start up protocols between trials should be the same. Fixed warm-up period for the vehicle, same time of day, and subject clothing should all be controlled where possible. If this is not the case, the experimenter should document fully all those external elements that may influence the responses of the subject.

If specific components are being compared, then the operating conditions and their order of presentation to subjects may be limited by practical issues, that is, changing glazing in a vehicle. The integration of subject numbers, measurements required and tests condition and available resources will influence the overall experimental design. But, the investigators should aim to be rigorous in their execution of the trial.

Figure 7.4 gives an example of a vehicle equipped for a field trial, it also shows a subject in a standardised clothing ensemble.

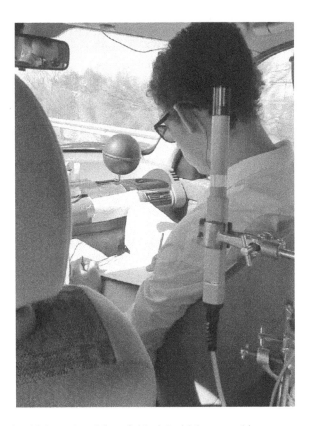

FIGURE 7.4 A vehicle equipped for a field trial with human subject.

TABLE 7.3
Desirable Thermal Environmental Conditions

	Summer[a]	Winter[b]
Air temperature	22–26.5°C	18–25°C
Mean air velocity	0.24 ms-1	0.21 ms-1
Relative humidity	40–60%	40–60%

[a] Assuming a Clo value of 0.5.
[b] Assuming a Clo value of 1.0.

7.3.4 GUIDELINES FOR DESIRABLE VEHICLE ENVIRONMENTAL CONDITIONS

As has been established, occupant environments are highly prone to the external climatic conditions. This makes it difficult to provide precise values for the internal environmental conditions. It is known that it is the interaction of the six parameters that will bring about the overall response from the occupant. It is often better to think of desirable ranges of conditions. Table 7.3 presents conditions for summer and winter conditions based around guidance in ISO 7730. These conditions should leave the vehicle occupant feeling between slightly cool and slightly warm.

With the scope for personal control within the vehicle, passengers should be able to 'fine tune' their environment to an acceptable level.

7.3.5 GENDER

There are no significant differences the preferred temperatures for males and females although it has been shown that females are most sensitive to deviations from thermo neutral temperatures (Fanger, 1970) However, there is some evidence of differences in the preferred comfort temperatures between males and females. Bedford (1936) showed that men preferred cooler temperatures than women, but subsequent studies have not found such a large difference between sexes (Fanger and Langkilde, 1975). Trends for females to be more sensitive to thermal deviations from thermal neutrality have been shown in several studies (Rohles and Nevins, 1971; Hodder et al., 1998). McIntyre (1980) suggested the female response to temperature change was faster than for males, with females getting hotter or colder significantly quicker than males. One of the factors that varies noticeably between the genders is clothing. Variations in clothing insulation within a given environment may vary significantly between male and females. This all suggests that females will be more prone to rapid changes within the thermal environment of a vehicle and minimising the time to reach acceptable thermal conditions would be of benefit.

7.3.6 GOALS FOR FUTURE

Improvements in occupant thermal comfort could be gained from decreasing warm-up and cool-down times for the passenger space to reach the desired thermal criteria.

The use of air flow is likely to remain the main system for controlling the environmental conditions. But it can be seen that there are potential benefits of localised heating and cooling to optimise passenger comfort.

Radiant temperature, especially the solar radiation component, remains a significant factor affecting the human thermal environment in the vehicle. The introduction of low transmission glazing can have benefits, particularly once journeys have commenced and the occupant space is controlled to a thermally acceptable level.

REFERENCES

ASHRAE. 1993. Physiological principles and thermal comfort, In *ASHRAE Fundamentals, SI edition*. Atlanta, US: ASHRAE.

Atkinson, W. J. 1987. Occupant comfort requirements for automotive air conditioning systems. *SAE* 860591.

Bedford, T. 1936. The warmth factor in comfort at work: A physiological study of heating and ventilation, *Industrial Health Research Board Report, no. 76*.

Brooks, J. E. and K. C. Parsons. 1999. An ergonomics investigation into human thermal comfort using an automobile seat heated with encapsulated carbonized fabric (ECF). *Ergonomics* 42 (5): 661–73.

Cengiz, T. G. and F. C. Babalık. 2007. An on-the-road experiment into the thermal comfort of car seats. *Applied Ergonomics* 38 (3) (5): 337–47.

Clark, R. P. 1981. Human skin temperature and convective heat loss. In *Studies in Environment Science, Bioengineering, Thermal Physiology and Comfort*. Vol. 10. Amsterdam: Elsevier Scientific Publishing Company.

Clark, R. P. and O. G. Edholm. 1985. *Man and His Thermal Environment*. London: Edward Arnold Ltd.

de Dear, R. and G. S. Brager. 2001. The adaptive model of thermal comfort and energy conservation in the built environment. *International Journal of Biometeorology* 45 (2): 100–108.

Fanger, P. O. 1970. *Thermal Comfort*. Copenhagen: Danish Technical Press.

Fanger, P. O. and G. Langkilde. 1975. Interindividual differences in ambient temperatures preferred by seated persons. *ASHRAE Transactions*, 81 (2): 140.

Fiala, D., K., J. Lomas and M. Stohrer. 1999. A computer model of human thermoregulation for a wide range of environmental conditions: The passive system. *Journal of Applied Physiology* 87 (5): 1957.

Fiala, D., K., J. Lomas and M. Stohrer. 2001. *Computer prediction of human thermoregulatory and temperature responses to a wide range of environmental conditions*. Vol. 45. Berlin/ Heidelberg: Springer.

Fung, N. and Parsons, K. C. 1996. Some investigations into the relationship between car seat cover materials and thermal comfort using human subjects, *Journal of Coated Fabrics*, 26, October.

Fung, W. 1997. How to improve thermal comfort of the car seat. Paper presented at *Proceedings of the 4th International Conference Comfort in the Automotive Industry, Recent Developments and Achievement*, Bologna, Italy.

Gagge, A. P. 1940. Standard operative temperature, a generalized temperature scale, applicable to direct and partitional calorimetry. *American Journal of Physiology*: 93.

Gagge, A. P., J. A. J. Stolwijk and J. D. Hardy. 1967. Comfort and thermal sensations and associated physiological responses at various ambient temperatures. *Environmental Research* 1 (1) (6): 1–20.

Gameiro da Silva, M. C. 2002. Measurements of comfort in vehicles. *Measurement Science and Technology* 13 (6): R41.

Givoni, B. 1976. *Man, Climate and Architecture*, 2nd ed. London: Applied Science.

Hodder, S. G., D. L. Loveday, K. C. Parsons, and A. H. Taki. 1998. Thermal comfort in chilled ceiling and displacement ventilation environments: Vertical radiant temperature asymmetry effects. *Energy and Buildings* 27 (2) (4): 167–73.

Hodder, S. G. and K. Parsons. 2007. The effects of solar radiation on thermal comfort. *International Journal of Biometeorology* 51 (3): 233–50.

Holmér, I. 2004. Thermal manikin history and applications. *European Journal of Applied Physiology* 92 (6): 614.

ISO 7243. 1989. *Hot Environments—Estimation of the Heat Stress on Working Man, Based on the WBGT-Index (Wet Bulb Globe Temperature)*. Geneva: International Standards Organization.

ISO 7726. 2001. *Ergonomics of the Thermal Environment—Instruments for Measuring Physical Quantities*. Geneva: International Standards Organization.

ISO 7730. 2005. *Ergonomics of the Thermal Environment—Analytical Determination and Interpretation of Thermal Comfort Using Calculation of the PMV and PPD Indices and Local Thermal Comfort Criteria*. Geneva: International Standards Organization.

ISO 8996. 2004. *Ergonomics of the Thermal Environment—Determination of Metabolic Rate*. Geneva: International Standards Organization.

ISO 9920. 2007. *Ergonomics of the Thermal Environment—Estimation of Thermal Insulation and Water Vapour Resistance of a Clothing Ensemble*. Geneva: International Standards Organization.

ISO 14505-2. 2006. *Evaluation of the Thermal Environment in Vehicles, Part 2: Determination of Equivalent Temperature*. Geneva: International Standards Organization.

ISO 14505-3. 2006. *Ergonomics of the Thermal Environment—Evaluation of the Thermal Environment in Vehicles, Part 3: Evaluation of Thermal Comfort Using Human Subjects*. Geneva: International Standards Organization.

Kerslake, D. McK. 1972. *The Stress of Hot Environments*. London: Cambridge University Press.

Madsen, T. L., B. W. Olesen and K. Read. 1986. A new method for evaluation of the thermal environment in automotive vehicles. *ASHRAE Trans.* 92: 38–54.

Madsen. T. L. 1994. Thermal effects of ventilated car seats. *International Journal of Industrial Ergonomics* 13 (3) (5): 253–8.

Matsunaga, K., F. Sudo, S. Tanabe and T. L. Madsen. 1993. Evaluation and measurement of thermal comfort in the vehicles with a new thermal manikin, *Society of Automotive Engineers,* 931958.

McIntyre, D. A. 1980. *Indoor Climate*. London: Applied Science Publishers Ltd.

McKinlay, A. F., F. Harlen and M. J. Whillock. 1988. *Hazards of Optical Radiation, a Guide to Sources, Uses and Safety,* IOP Publishing Ltd., London.

Mezrhab, A. and M. Bouzidi. 2006. Computation of thermal comfort inside a passenger car compartment. *Applied Thermal Engineering* 26 (14–15) (10): 1697–1704.

Nielsen, B. 1990. Solar heat load: Heat balance during exercise in clothed subjects. *European Journal of Applied Physiology and Occupational Physiology* 60 (6): 452–6.

Nilsson, H., I. Holmer, M. Bohm and O. Noren. 1997. Equivilant temperature and thermal sensation—Comparison with subjective measurements. Paper presented at *Proceedings of the 4th International Conference Comfort in the Automotive Industry, Recent Developments and Achievement*. Bologna, Italy.

Oi, Ha, K. Tabata, Y. Naka, A. Takeda and Y. Tochihara. 2012. Effects of heated seats in vehicles on thermal comfort during the initial warm-up period. *Applied Ergonomics* 43 (2) (3): 360–7.

Parsons, K. C. 2003. *Human Thermal Environments: The Effects of Hot, Moderate, and Cold Environments on Human Health, Comfort, and Performance*, 2nd ed. London: Taylor & Francis.

Parsons, K. C. and D. Entwistle. 1993. An investigation into the thermal of motor automobile drivers. Paper presented at *Proceedings of the Ergonomics Society Conference.*

Rohles, F. H. and R. G. Nevins. 1971. The nature of thermal comfort in sedentary man, *ASHRAE Journal 2191 RP-43.*

Rohles, F. H. and S. B. Wallis. 1979. Comfort criteria for air-conditioned automotive vehicles. *Society of Automotive Engineers,* 790122.

Roller, W. L. and R. F. Goldman. 1968. Prediction of solar heat load on man. *Journal of Applied Physiology* 24: 7817.

Stolwijk, J. A. J. and Hardy, J. D. 1977. Control of body temperature, in *Handbook of Physiology, Section 9: Reaction to Environmental Agents.* Bethesda, Maryland: American Physiological Society, pp. 45–68.

Temming, J. 1980. *Comfort Requirements for Heating, Ventilation and Air-Conditioning in Motor Vehicles. Human Factors in Transport Research,* Eds. Oborne, D. J. and Lewis, J. A., Vol. 2. Academic Press: New York.

Wyon, D. P., C. Tennstedt, I. Lundgren and S. Larsson. 1985. A new method for the detailed assessment of human heat balance in vehicles—Volvo's thermal manikin, VOLTMAN. *SAE Paper,* 850042.

Yaglou, C. P. 1927. Temperature, humidity and air movement in industries: The effective temperature index. *Journal of Industrial Hygiene* 9: 297–309.

Zhang, Y, F., Wyon, D. P., Fang, L. and Melikov, A. K. 2007. The influence of heated or cooled seats on the acceptable ambient temperature range. *Ergonomics* 50 (4).

Zimmy, K., H. Zenker, S. Doemoek and M. Ellinger. 1999. Comparison between measured and computer-simulated t_{eq}. Paper presented at *Proceedings of the 6th International ATA Conference,* Florence, Italy.

8 Driving Posture and Healthy Design

Diane Elizabeth Gyi
Loughborough University, UK

CONTENTS

8.1 INTRODUCTION

So, why is driving sometimes a pain? Driving as a task involves prolonged sitting, a static and constrained posture, vibration and muscular effort (from steering, braking, reversing etc.), all loading the spine to varying degrees and any of which individually could lead to musculoskeletal symptoms. For successful vehicle seat design and good ergonomics, some knowledge of the anatomy, physiology and biomechanics of the human seated posture is required. In the view of the author, it is important to understand 'the human body' and the impact that design decisions will have on behaviours that affect driver comfort and health. As a contribution to this goal, this chapter provides an overview of 'the basics' of the human seated posture, why driving is a pain, and principles of good ergonomics in the driving task. Although it mainly focuses on car driving, some items are of relevance to other vehicles.

8.2 THE SEATED POSTURE AND DRIVING

Traditional human factors texts clearly document the anatomical and physiological factors involved in sitting. The efficiency of any posture from a simple biomechanics viewpoint can be determined by the degree to which it loads the skeleton and postural muscles. Postural stress is a result of gravitational (and other forces) acting

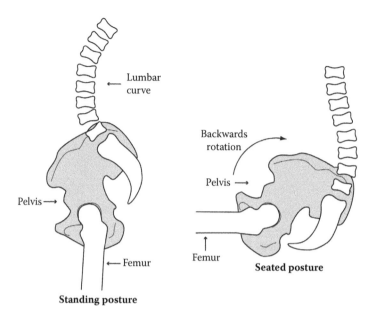

FIGURE 8.1   Rotation of the pelvis when changing from standing to a seated posture.

on the body and the forces required by muscle activity to maintain any particular posture (Troup, 1978). In fact, the muscular effort required for sitting is greater
than that for standing as shown by Nachemson, Andersson, and Schultz (1986) in
an experimental study using electromyography. In another early study, Andersson et
al. (1974), using a transducer mounted in a hypodermic needle also found that intra-
discal pressure in the spine was 40% higher in sitting than in standing. These early
studies have implications for the modern driving task.

When changing from standing to a seated posture, backwards rotation of the pelvis flattens the curve of the lumbar spine and changes its shape (Figure 8.1). This
increases pressure on the posterior part of the inter-vertebral discs and within the
nucleus itself making it vulnerable to damage (Figure 8.2).

The lumbar curve could be actively maintained by contraction of the muscles in
the back (e.g., latissimus dorsi and the sacrospinalis) but this is very tiring. So, when
sitting on a seat with a backrest, the pelvis will rotate backwards until the person's
back comes into contact with the support. In a well designed seat, the weight of the
trunk is taken by the backrest, the muscles are then relaxed and the curve of the
lumbar spine is supported. Conversely, in a poorly designed seat, the lumbar curve is
flattened (a loss of lordosis) increasing pressure within the discs (Figure 8.2), strain-
ing the spinal ligaments and gluteal muscles and increasing the thoracic c-shaped
curve in the upper spine (increase in kyphosis). In a car, this slouched posture could
be exacerbated by design elements, such as low headroom space or a seat cushion
length, which is too long. So, although this slouching reduces the need for muscular
effort in the trunk, it increases disc loading.

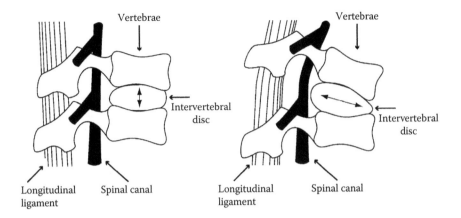

FIGURE 8.2   Diagrammatic representation of vertebrae and discs.

Varying posture during the working day results in changes in disc pressure, which is beneficial for the vertebral discs which have no blood supply of their own. Krämer (1973) demonstrated that compression of the vertebral disc causes diffusion of fluid from the interior to exterior. With a reduction in the compressive force, this process is reversed and tissue fluid diffuses back in bringing essential nutrients with it. To keep discs well nourished and in good condition, ideally they need to be subjected to frequent changes in posture. Facilitating this through driver behaviour supported by good design is important for spine health.

New research is also looking into the importance of 'loading history' on the biomechanics of the spine and indicates that following prolonged sitting, manual handling tasks should be avoided and care should be taken until the disc has had chance to recover (McGill, 2007). McGill and Brown (1992) cite up to 40 minutes is needed, although approximately 50% recovery is achieved after 2 minutes of standing. From the literature, the advice is to never lift heavy items after prolonged sitting, which has implications for drivers lifting items from the car boot, emergency service workers and therefore the design of the vehicle. Good car seat design supporting the lumbar curve and avoiding the flexed 'c-shaped' curve remains a good strategy for these workers. In fact, following prolonged sitting activities such as driving, time should be spent standing and ideally walking around before manual handling out of the car (McGill, 2007). In practice, this could be easily achieved by, for example, walking around to make a phone call, getting a drink, finding out exactly where meeting place is and so on, before unloading items.

An element of static muscle work is also present in the seated driving posture. The contraction of muscle tissue leads to compression of the blood vessels thereby reducing its blood supply and disrupting nutrient delivery and metabolite removal (e.g., lactic acid). It is the accumulation of these metabolites that produces localised muscle fatigue and acute discomfort. Delaying or preventing these undesirable effects could be achieved by periodically relieving the muscles of their activity, that is, varying posture, for which there is little opportunity during driving. In fact, Callaghan and McGill (2001) suggest that there is no single ideal seated posture and that a variable posture is the best strategy to minimise muscle tissue overload. Consequently, even the most comfortable posture can be fatiguing over time: there is no 'ideal sitting posture' but the ideal scenario is a seat which safely allows variable postures. Further research is needed in this area.

Whilst the vehicle is in motion, different forces act on the body (acceleration and deceleration, lateral and vertical movements) and when the feet are actively being used, the degree to which they can support and stabilise the body is limited. The task of driving also involves muscular effort; steering, braking, clutch work, using the handbrake, reversing and so forth, all of which load the spine to varying degrees. For example, psoas major, a powerful hip flexor is used each time a foot is lifted onto a pedal. Adverse postures are also needed for driving such as those required for reversing, involving the extensors and rotators of the cervical and thoracic spine compressing the vertebral discs. There is no evidence in the literature that suggests that the muscular effort from driving itself leads to musculoskeletal symptoms, but many authors agree that the relationship between driving and such symptoms does warrant further investigation.

During driving, the postures adopted depend on arrangement of the pedals, steering wheel and the seat itself in the driving workspace together with the visual demands of the set-up. Environmental influences such as no support for the feet, low-friction seating material or poor steering wheel height can also all create additional muscle work. Poor design forces the adoption of awkward and inefficient working postures that ultimately lead to discomfort, pain and chronic disability if adverse conditions persist.

8.3   DRIVING AND MUSCULOSKELETAL HEALTH

It is well documented that the constrained and relatively static postures imposed by seated work are risk factors for musculoskeletal ill health. However, epidemiological studies examining the relationship between car driving and back pain or other musculoskeletal disorders are relatively few, which is perhaps indicative of the methodological difficulties of conducting such studies. Kelsey and Hardy (1975) found a link between driving occupations and acute herniated lumbar inter-vertebral discs in a matched-pairs study of patients who attended x-ray departments from 1971 to 1973. They found that comparing cases to matched controls, if the case had a job where they spent more than half their time in a motor vehicle, they were nearly three times more likely to develop an acute herniated lumbar disc. The study was not designed to look specifically at driving, yet it appeared as a factor reducing the likelihood that this association could have occurred by chance. The vast majority of individuals reporting back pain do not require hospital intervention and therefore these results could be the 'tip of the iceberg'. Pietri et al. (1992) carried out an extensive interview study of commercial travellers in France. Respondents with low back pain were compared to those without, and the risks of low back pain were significantly associated with driving more than 20 hours a week. There is evidence that those who drive as part of their job are particularly vulnerable to musculoskeletal symptoms, with those who drive more than 4 hours/day, more than 6 times more likely to take sick leave with low back symptoms than those with lower mileage (Porter and Gyi,

2002). Indeed a high prevalence of musculoskeletal symptoms have been found in driving professions, for example, in a recent study of pharmaceutical sales representatives (Sang, Gyi and Haslam, 2010), 84% indicated musculoskeletal symptoms in the last 12 months, the most prevalent were in the lower back (54%), the neck (46%), and the shoulders (45%). Prevalence rates are similar to other studies in Japan (Sakakibara, Kasai and Uchida, 2006) and in Turkey (Tander, Canbaz and Canturk et al., 2007). As well as driving, drivers in the study by Sang et al. (2010), reported using the car as a mobile office (e.g., laptop and mobile phone use, writing notes) and manual handling items from the car. This has also been noted by other authors (East and Flyte, 1998). Whilst training to raise driver awareness of the risks from these activities is important, design elements of the vehicle (e.g., boot design, storage) will also play a part. Stuckey, Lamontagne and Sim (2007) suggests that driving is often considered incidental to an individual's job and that health risks are not considered by either the driver themselves or their managers. Raising awareness of the risks and the importance of selecting a suitable vehicle that meets their individual and task/ work related needs is paramount.

8.4   THE OLDER DRIVER

The World Health Organization has predicted that by 2025, one-tenth of the world population will be over 65 years old. The ageing process combined with the effects of previous injury or disease will have an effect on anthropometry and physical capabilities including those required of the driving task. This will affect variability within individuals over a lifetime, for example, longitudinal studies show that stature (and associated measures), decrease with age particularly in women largely due to muscle atrophy, compression and thinning of the inter-vertebral discs. Body size and girth also have been shown to decrease in later years at about 50 years in men and 60 years in women (Pheasant and Haslegrave, 2006). There is also a reduction in range of motion and flexibility of joints with increasing age (Steenbekkers, 1998) and age delays the recovery of spinal tissues (Twommey and Taylor, 1982).

Smith, Meshkati and Robertson (1993) have put together an excellent overview of the literature concerning the older driver and passenger together with considerations for automotive design. They remind the reader, that older people show the greatest individual variability of any age cohort and that relying on chronological age as predictors of physical and behavioural aspects is likely to be unreliable. For example, in an anthropometry study of 750 participants it was found that although force exertion decreases with age, the differences between males and females are much larger than those between age groupings (Voorbij and Steenbekkers, 1998). There are many good texts regarding anthropometric data which are useful for design (e.g., Pheasant and Haslegrave, 2006; Steinbekkers and Van Beijsterveldt, 1998). However, data are limited regarding specific dynamic and functional anthropometric measurements relevant to vehicle design, for example ingress into and egress from the vehicle, reach to seat belts, reach to adjust mirrors, opening car doors, opening car boots, postures for reversing and operating seat adjustment controls. Studies in this area are important with the changing demographics and design to include older drivers of the future.

8.5 DESIGN ERGONOMICS, COMFORT AND DRIVING

A good overview of the science of vehicle seat comfort is given by Kolich (2008). He indicates that seat comfort is multi-dimensional and based on his experience in the automotive industry provides sensible and practical insights. The vehicle package itself (e.g., headroom, legroom) defines the workspace, so the same seat placed in two different vehicles is very likely to receive different comfort ratings. This is important for researchers to consider when conducting experimental studies on automotive seats. Proportions between individuals vary but even individuals with similar anthropometry may have different preferences in terms of driving posture due to factors such as joint flexibility and different driving behaviours. He summarises that factors affecting subjective perceptions of comfort include:

1. Individual factors, for example, demographics, anthropometry, culture and posture.
2. Vehicle/driving work package, for example, seat height/eye point, pedal/ steering wheel position and head/knee room.
3. Seat factors, for example, stiffness, geometry (dimensions), contour (shape), breathability and styling.
4. Social factors, for example, vehicle nameplate, brand and cost of the vehicle.

It remains true that an important contribution of ergonomics to the vehicle design process is information concerning occupant shapes and sizes, for which there are several good texts, for example, Henry Dreyfuss Associates (2002) and Pheasant and Haslegrave (2006).

Based on our research, generally the greater the number of adjustable features in the vehicle, the more likely it is that a range of comfortable postures and a good 'fit' can be achieved by the driver (see Figure 8.3). For example, drivers should look for independent seat height and tilt adjustment, backrest adjustment, lumbar support adjustment (up/down, in/out) and even seat length adjustment. From a designer's perspective, this means that seat depth should not exceed buttock popliteal length of a small user (1st–5th percentile females of the relevant population). A seat depth which is too long will also deprive the user of the benefits of the backrest, giving the person no choice but to lean back with the lumbar curve essentially unsupported. Drivers should also check that the backrest provides support along the length of the back and that the head restraint can be correctly positioned. In general, the higher the backrest contoured to the shape of the spine, the better the postural support it will give. According to Pheasant and Haslegrave (1996), the midpoint of the curvature should be about 230 mm above the seat surface. An adjustable lumbar support can be an advantage provided it can be shaped to 'fit' and support the lower back with no pressure points or gaps. The curve in a lumbar support should not be too excessive as a curve which is too pronounced may be worse than none at all. Full adjustability of the steering wheel is also desirable and drivers should check for clearance of the legs/ arms when operating the vehicle and using the controls (e.g., pedals, hand brake). Drivers should also be aware of coping strategies, such as limited headroom which can force a reclined posture and undesirable excessive bending of the head and neck.

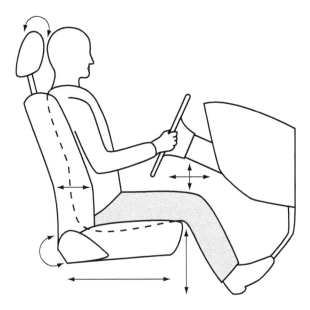

FIGURE 8.3 Diagram to show optimum seated posture and adjustments.

This may also make reach to the steering wheel difficult and lead to forced extension of the arms. Research by Porter and Gyi (1998) provides guidelines for optimum postural comfort. They also indicate that many drivers particularly those at the extremes of anthropometric dimensions (e.g., long legs/arms/sitting height, short legs/arms/sitting height) may need to compromise their preferred posture to fit many vehicles on the market today. Figure 8.4 shows a useful checklist of optimal requirements which can be used when selecting a vehicle from an ergonomics and health perspective to avoid some of these problems. These also act as a useful checklist for human factors specialists working in or with the automotive industry.

8.6 SUMMARY

Prolonged exposure to driving cars is a risk factor for low back and other musculoskeletal symptoms. Poor design of the driving workstation forces the adoption of awkward and inefficient postures that ultimately contribute to musculoskeletal discomfort. Design teams should consider the design parameters that support good 'fit', encourage the adoption of good postures and driver behaviours, whilst taking into account the principles of limiting postural stress. A key element in facilitating this is through more thoughtful and effective design.

ACKNOWLEDGMENT

I would like to thank Dr Elaine Yolande Williams for her help with the graphics.

DRIVING
ergonomics

Car selection checklist

Test drive

	Yes
Have you test driven the car?	☐
Has the test drive lasted at least two hours?	☐

The seat

	Yes
Does the driver's seat have independent tilt adjustment?	☐
Does the driver's seat have independent height adjustment?	☐
Are the seat adjustment controls easy to use?	☐
When you have adjusted your seat are you able to reach the pedals without stretching?	☐
Does the back rest reach your shoulders?	☐
Does the back rest provide support along the length of your back?	☐
Is your lumbar curve supported without any points of pressure or gaps?	☐
Do you have enough leg and head room?	☐
Does the seat length put pressure on the back of your knees or calves?	☐
Is the head restraint positioned close to your head?	☐
Is the head restraint height near the top of your head?	☐
Are you able to get in and out of the car easily?	☐

The steering wheel

	Yes
Is the steering wheel centrally located?	☐
Does the steering wheel have the following adjustment features:	☐
In/out?	
Up/down?	☐
Tilt ?	☐
Do you have full view of the display panel?	☐

The pedals

	Yes
Are the pedals centrally positioned?	☐
Is there a left foot rest?	☐
Is there plenty of room for you to rest your left foot?	☐

The boot

	Yes
Does the car have adequate boot space for you to store equipment or other items?	☐
Are you able to easily access your boot without obstruction, for example, from the parcel shelf, or the boot lid itself?	☐
Does the boot have a low or flat sill height?	☐
Does the boot have a narrow sill depth?	☐
Does the boot have handles to facilitate closing?	☐

FIGURE 8.4 Car selection checklist. (From www.drivingergonomics.com. ©Copyright Loughborough University.)

REFERENCES

Andersson B. J. G., Ortengren R, Nachemson A. and Elfstrom G., 1974. Lumbar disc pressure and myoelectric back muscle activity during sitting IV. Studies on car drivers seat. *Scandinavian Journal of Rehabilitation Medicine*, 6, 128–133.

Callaghan J. P. and McGill S. M., 2001. Low back joint loading and kinematics during standing and unsupported sitting. *Ergonomics*, 44, 373–381.

Eost C. and Flyte M. G., 1998. An investigation into the use of the car as a mobile office. *Applied Ergonomics*, 29, 283–388.

Henry Dreyfuss Associates, 2002. *The Measure of Man and Woman: Human Factors in Design*, rev. ed. New York: John Wiley & Sons.

Kelsey J. L. and Hardy J., 1975. Driving of motor vehicles as a risk factor for acute herniated lumbar vertebral disc. *Journal of Chronic Disorders*, 28, 37–50.

Krämer J., 1973. Biomechanische veränderungen im lumbalen bewegungssegment. In E. Grandjean, (ed.) *Fitting the Task to the Man*. London: Taylor & Francis, p. 57.

Kolich M., 2008. A conceptual framework proposed to formalise the scientific investigation of automobile seat comfort. *Applied Ergonomics*, 39, 15–27.

McGill S., 2007. *Low Back Disorders: Evidence-Based Prevention and Rehabilitation*, second edition. University of Waterloo, Canada: Human Kinetics.

McGill S. M. and Brown S., 1992. Creep response of the lumbar spine to prolonged lumbar flexion. *Clinical Biomechanics*, 7, 43.

Nachemson A. L. G., Andersson B. J. and Schultz A. B., 1986. Valsalva maneuver biomechanics. Effects on lumbar trunk loads of elevated intra-abdominal pressures. *Spine*, 11, 476–479.

Pheasant S. and Haslegrave C. M., 2006. *Bodyspace: Anthropometry, Ergonomics and the Design of Work*. London: Taylor & Francis.

Pietri F., LeClerk A., Boitel L., Chastang J., Morcet J. and Blondet M., 1992. Low back pain in commercial travellers. *Scandinavian Journal of Work and Environmental Health*, 18, 52–58.

Porter J. M. and Gyi D. E., 1998. Exploring the optimum posture for driver comfort. *International Journal of Vehicle Design*, 19, 255–265.

Porter J. M. and Gyi D. E., 2002. The prevalence of musculoskeletal troubles among car drivers. *Occupational Medicine*, 52, 4–12.

Sakakibara T., Kasai Y. and Uchida A., 2006. Effects of driving on low back pain. *Occupational Medicine*, 56, 494–496.

Sang K., Gyi D. E. and Haslam C., 2010. Musculoskeletal symptoms in pharmaceutical sales representatives. *Occupational Medicine*, 60, 108–114.

Smith B. D., Meshkati N. and Robertson M. M., 1993. The older driver and passenger. In B. Peacock and W. Karwowski, (eds.) *Automotive Ergonomics*. London: Taylor & Francis, pp. 453–467.

Steenbekkers L. P. A., 1998. Ranges of movement of joints. In L. P. A Steenbekkers and C. E. M Van Beijsterveldt, (eds.) *Design-Revelevant Characteristics of Ageing Users*. Delft: Delft University Press, pp. 60–68.

Steenbekkers L. P. A. and Van Beijsterveldt C. E. M., 1998. *Design-Revelevant Characteristics of Ageing Users*. Delft: Delft University Press.

Stuckey R., Lamontagne A. D. and Sim M., 2007. Working in light vehicles—A review and conceptual module for health and safety. *Accident Analysis Prevention*, 39, 1006–1014.

Tander B., Canbaz S., Canturk F. and Peksen Y., 2007. Work related musculoskeletal problems among pharmaceutical sales representatives in Samsun, Turkey. *Journal of Musculoskeletal Rehabilitation*, 20, 21–27.

Troup J. D. B, 1978. Drivers back pain and its prevention. *Applied Ergonomics*, 9, 207–214.

Twommey L. and Taylor J., 1982. Flexion creep deformation and hysteresis the lumbar vertebral column. *Spine*, 7, 116–122.

Voorbij A. I. M. and Steenbekkers L. P. A., 1998. Exertion of force. In L. P. A Steenbekkers and C. E. M van Beijsterveldt, (eds.) *Design-Revelevant Characteristics of Ageing Users*. Delft: Delft University Press, pp. 48–59.

9 The Essential Realism of Driving Simulators for Research and Training

Andrew Parkes
Transport Research Laboratory, UK

Following the lead set by the aviation industry, simulators have been developed for both research and training for road vehicle drivers. There are cheap ones, expensive ones, large ones and small ones; many are operated by experienced and knowledgeable staff, some are not. In short, a wide variety of systems exist that can be described as a driving simulator and it can be difficult to distinguish those that might be useful and promote realistic behaviour and effective learning experiences. This chapter attempts to trace some of the factors driving the current development of simulators and also looks at ways in which they could be categorised in terms of their function rather than just in terms of their cost or their technical performance.

Figure 9.1 shows an early training system for aircraft. This intriguing device allowed the potential pilot to be moved by his supporters in several planes of movement and apparently learn some elements of the skill of maintaining orientation to the horizon. There would be elements of strategy and tactics involved, but the emphasis would have been on the control actions of the pilot and the benefits of exposure to the physical sensations of being tipped back and forth accordingly. How effective in terms of transfer of training is difficult to determine, but the principle of simulating aspects of the complex and dangerous task, and allowing the pilot to practice in a safe, benign and repeatable environment was certainly soon established. As with many technologies, progress was accelerated by the demands of times of war and as the aircraft became more sophisticated and important information needed to be regularly monitored by the pilot, there emerged devices that would be recognised by the simulation training community today.

An interesting example is given in the Figure 9.2 below. The wartime task of producing a new trainer was given in the United States to Bell Telephone Laboratories

FIGURE 9.1   Pilot task simulation.

who produced an operational flight trainer for the Navy's PBM-3 aircraft. This device, completed in 1943, consisted of a replica of the PBM front fuselage and cockpit, complete with controls, instrumentation and all auxiliary equipment, together with an electronic computing device to solve the flight equations. The simulator had no motion system, visual system or variable control loading. A total of 32 of these electronic flight trainers, for seven types of aeroplane, were built by Bell and the Western Electric Company during the war years. The PBM-3 was possibly the first operational flight trainer that attempted to simulate the aerodynamic characteristics of a specific aircraft. The salient point here is that effective and efficient training was delivered on something that did not (could not at that time) attempt to simulate the whole task for the pilot. There was no motion or rendering of the visual landscape. The system was produced to provide training on only part of the pilot's task.

   This notion of partial simulation and the whole question of how realistic and complete an experience needs to be in order for it to be useful from either a training or research perspective underlie and complicate current debate. It is clear that the aviation industry, both military and civilian, believes in and derives great benefit from advanced simulation techniques. Pilots are trained on simulators as part of their required curriculum. Indeed, for certain aircraft all the training hours required can be accomplished in an appropriate accredited and certified simulator. Advanced aircraft

FIGURE 9.2   PBM-3 aircraft simulator.

FIGURE 9.3   Advanced modern aircraft simulation.

simulators (Figure 9.3), though very expensive, are seen as cost effective when compared to the cost of access to the real airplane or to the cost of damage or loss of the aircraft due to mistakes made during the training programme. Much of the value is seen in the ability to recreate emergency situations, such as equipment failure or extreme weather conditions, that would be impracticable or unsafe to attempt in real life.

The case for driving simulators is less clear. Lee (2004) gives examples of early training systems, including the Iowa State driving simulator from 1958, that linked a vehicle cab mock-up to a scaled physical terrain model, allowing the driver to control actions of models in a rudimentary road layout. Such a system was not intended to develop the learning of the control actions of a vehicle but was aimed at allowing the student to learn something about rules of the road and anticipatory behavior. Since then there have been many different technical innovations including video of real scenes and more recently, computer-generated environments that have led to immersive experiences that can be very similar to the sensations of driving a real vehicle. Military interest in simulation has been strong and has supported the development of many top-end systems, while in recent years the mass-market computer game industry has included a host of action and racing titles that have pushed the degree of apparent realism forward.

In terms of serious applications of driving research and training, there have been two main strands of development. Research simulators have been required for applied experimentation that is either too costly or too dangerous to conduct on the

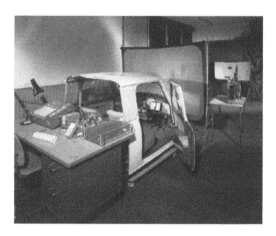

FIGURE 9.4   Early film-based simulator at TRL circa 1970.

real road; for example, driver distraction and impairment studies, or investigations into new highway design options. Training simulators have been developed that have either targeted the absolute novice driver and been designed to allow them to learn the most basic control concepts within a benign environment, or have been targeted at professional drivers (e.g., the police) for more tactical training of responses in emergency or high workload conditions.

Figure 9.4 shows as early example of a research simulator based at the Transport Research Laboratory (TRL) in the UK. The simulator required the driver to sit in a real vehicle cab and view scenes filmed from real road situations. The task was interactive to the extent that the driver was required to rate the apparent level of risk in the scene continuously.

The drivers had controls to indicate their perception of the emerging dangers in the road scene, but did not control the vehicle in any way (Figure 9.5). They were passive observers of the visual image; there was no sound, no movement of the

FIGURE 9.5   Real road scene projected to the driver.

FIGURE 9.6   Group training circa 1968 in Sweden.

vehicle cab and no way of engaging directly with the images in terms of the position of the vehicle. As such, it was not like driving a real car. It was however a valuable way of learning about hazard perception and has led directly to the establishment of a test that is now required as part of the licence acquisition process for all novice drivers in the UK.

Use of basic systems for training car drivers in some of the elements of control operation or road tactics is well established. Figure 9.6 above shows mock vehicle cabs in a classroom setting in Sweden where all the students are responding to the same projected image of a road scene.

There would be no realistic feel to the vehicle controls and the group participation in choreographed rule-based behaviours would not produce an immersive driving experience, but at that time at least, it was considered a valuable addition to on-road training, particularly for those drivers toward the start of their learning schedule.

Since the 1970s, technology and in particular, computer technology, has advanced rapidly. The ability to provide reasonably realistic physical sensations to the driver, from movement of the car shell and the feel of the controls, coupled to the provision of computer-generated images has led to a substantial growth in the use of simulation for both research and training (Carsten and Jamson, 2011).

Many of the national road safety research institutes and an increasing number of universities and training organisations now operate driving simulators as a part of their mainstream activities (Figure 9.7). At present, this is an unregulated area with complete freedom of entry to the training marketplace or to the scientific arena. Given that it is difficult to know, just by looking at a simulator, exactly what it can do, or how accurately it can do it, there is increasing demand from both the scientific and training communities for some means of categorising or accrediting simulators for different purposes.

Many of the most advanced driving simulators in North America and Europe are currently operated as truck driver training systems and it is for this community that

FIGURE 9.7   TRL interactive simulator circa 2006.

the first European-wide policy has been established. Progress, however, has been slow. It is interesting to note a 1996 report of the US Federal Highway Administration (FHWA) that detailed a scoping study on commercial motor vehicle driving simulator technology. It cited an earlier 1991 special issue of *Heavy Duty Trucking* that claimed:

> Cost-effective training simulators are becoming technologically possible—there have been astounding leaps in computer graphics and realism—at the same time the driver shortage and the Commercial Driver License (CDL) are forcing the trucking industry to seek more effective methods for driver training, selection and screening.

Some outside the industry might view it as surprising that, given the size of the trucking industry in the US and Europe, and the optimism displayed over twenty years ago, there are still relatively few (compared to the size of the industry) commercial truck simulators in existence, and little consensus on the content of any curriculum delivery. Indeed, the intervening period since the US FHWA study has seen continued technological development in simulators, particularly in visual database rendering, but very patchy uptake and development of simulation facilities for commercial truck driver training (Figure 9.8).

From a world-wide perspective, a clear lead has been taken by France and the Netherlands, but even in those countries there is neither the capacity to introduce simulation components to all truck drivers undergoing current training, nor to satisfy any potential increase in demand. There appear to be four fundamental reasons for the relatively slow adoption of simulation as a key component of professional truck driver training:

A lack of easily accessible documented evidence showing a clear benefit of simulation training over traditional on-road and test track methods.

FIGURE 9.8   Full mission truck simulator with advanced motion system.

A concern over the economics of providing high technology facilities and the
   attendant high costs of entry to the area.
A concern from the drivers that such training will be additional to, rather than
   replace parts of, the current requirements.
Some people get ill in simulators.

One might conclude that to date there has been a rather hard-to-identify *car-rot*, and a complete absence of any *stick* to encourage widespread development and
uptake of synthetic (driving simulator) training. The commercial truck sector is
very different to military ground vehicle, or aviation, sectors, where the presence
of cost–benefit models, accreditation and certification bodies, and agreed curricula
are evident. There has been a general assumption that simulators (or more correctly,
synthetic trainers) will probably eventually become widespread as computer costs
come down and power increases, but the freight industry and the driver training
industry is so fragmented in Europe, there is little to encourage early adopters of
the technology.

The picture, in Europe at least, may soon change. The European Commission
Directive on Training for Professional Drivers (EU Commission, 2001) adopted
by European Parliament in April 2003 stipulated that all persons wishing to drive
large goods vehicles (LGVs) in excess of 7.5 tonnes in a professional capacity, would
have to undergo training for, and obtain, a vocational Certificate of Professional
Competence (CPC) further to the LGV licence. The Directive provides a framework
for licence acquisition, testing and further skills development.

The total length of *full basic training* in the Directive is 420 hours (12 weeks of
35 hours each). For *minimum basic training*, this would be 280 hours. Each trainee
driver must drive for at least 20 hours individually in a vehicle of the category concerned. This Directive is of paramount importance to the training and simulation
industries, because for the first time, explicit reference is made to simulators for both
training and testing.

Each driver may drive for a maximum of 8 hours of the 20 hours of individual
training:

> ... on special terrain or on top-of-the-range simulators so as to assess training in rational driving based on safety regulations, in particular with regard to vehicle handling in different road conditions and the way they change with different atmospheric conditions and the time of day or night. (European Parliament, 2003, p. 24)

This wording does not go so far as to say that training *should* include simulation, nor that the time devoted to such training *should* be 8 hours, nor does it *recommend* simulation; but for the first time, it allows the possibility.

The Directive goes even further. It opens the way for simulation to play a part in the practical element of the driving test. It states that the basic elements of the practical test must have a duration of at least 90 minutes. This practical test may be supplemented by an assessment taking place on special terrain or on a top-of-the-range simulator.

> The duration of this optional test is not fixed. Should the driver undergo such a test, its duration may be deducted from the 90 minutes .... but the time deducted may not exceed 30 minutes. (European Parliament, 2003, p. 25)

So, simulation is seen as a viable medium for testing and early skills development for novice drivers. However, the training Directive is also concerned with the skill set of existing experienced drivers. A driver who has obtained his or her licence must undergo 35 hours of continuous training every 5 years.

> .... Such periodic training may be provided, in part, on top-of-the-range simulators. (European Parliament, 2003, p. 27)

The current wording poses some problems, for as yet, there is no satisfactory consensus view on the definition of *top-of-the-range*. It begs the question; who will be the arbiter and monitor of such a distinction?

According to the Directive, basic vocational training is divided into three areas:

Advanced training in rational driving based on safety rules.
Compliance with regulations.
Health, safety, service and logistics.

In addition, there are other areas of direct relevance to possible simulator training. These relate to:

Road traffic regulations.
Ergonomic principles.
Behaviour in an emergency situation.

This shows where simulation, and synthetic training in general, could provide a valuable role, but it does not *prescribe* exactly which elements may be suitable, nor *proscribe* those that are unsuitable. The introduction of compulsory basic, and continuous training, will require a large increase in capacity in the training industry. As the industry expands, there is a general expectation that simulation will become more common, and could eventually be a core component of the curricula. However, it could be a mistake to assume that simply because simulators are widespread, successful, and necessary in aviation or military ground vehicle applications, that they will be similarly well accepted and suitable for truck driver training and hence, later become routine components of all driver training programmes.

The review by Williges, Roscoe and Williges (1973), pondered the then 50-year history of flight simulation, and concluded that '.... many issues concerning ground based flight simulators and trainers remain unanswered'. Many concerns remain in aviation, and most remain to be addressed at all in a systematic fashion for road vehicles.

The possible benefits of simulation are clear. There is potential for: control of the training environment, repeatability of specific combinations of features, objective performance scoring, cost reduction and consistent online tutorial delivery. The training environment can also be more effective than the real world due to the ability to remove unessential elements from any particular scenario; and safer, due to the lack of physical risk, no matter how catastrophic the performance failure.

However, potential operators of training simulators need to know the following:

What can they really do?
How much will they cost?
What new skills will trainers need?
How will they be accredited?
How should simulators be used within a wider curriculum?

A similar set of questions could be posed by the research community. The problem at present is that whilst there are several convincing high-fidelity truck simulation systems available, there are very few answers available to the last of the questions above. There is little known in relation to truck driving, and little that is directly transferable from aviation, that can inform discussion of what should be delivered in a simulation training package, nor how the costs and benefits might compare to real road training. Information exists (e.g., Parkes, 2003, 2005; Reed and Parkes, 2005; Parkes and Reed, 2005b, 2006), but at present it is limited to a small number of systems.

Figure 9.9 below attempts to demonstrate part of the current dilemma. In an ideal world, we might hope there would be a clear linear relationship between the cost of a particular simulation system and the value of the training transfer that could be derived (line A). In reality, the relationship is likely to be less than straightforward.

There is certainly a strong suspicion it may become increasingly expensive to add fidelity as one moves along the line (B), and added expenditure may result in diminishing returns on investment. In practice, there are many go/no-go decisions to be made in simulator specification, and so a step function (line C) may be more realistic. Decisions such as whether to include a motion system, or to include multiple channel projection systems, or to include sophisticated three-dimension sound rendering, all require jumps in technology provision that have substantial cost implications.

The examples above show that systems can differ in terms of the *degree* of task simulation, from particular parts of the driving task (e.g., hazard perception, or use of certain controls) to recreations of the whole driving experience. However, for any

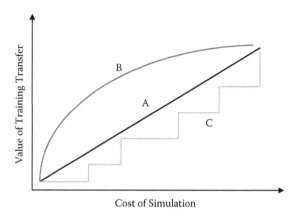

FIGURE 9.9   Models of fidelity versus cost of simulation.

given *degree* of simulation, there are also concerns over the *fidelity* or *validity* of the system. It is also possible that different types of validity are of particular importance for training or for research.

Validity has been thought of in various terms (e.g., Witmer and Singer, 1998; Kalawski, 2000; Godley, Triggs and Fildes, 2002; Kaptein, Theeuwes and Van der Horst, 2007) and though the list below is not comprehensive, it serves to highlight some of the important concepts:

Face validity.
Performance validity.
Construct validity.
Content validity.
Objective validity.
Relative validity.

Face validity refers to the initial look and feel of a system and how similar it is to the vehicle and task it simulates. Performance validity refers to the characteristics of the vehicle replicated and can be measured directly (e.g., acceleration, braking, sound profiles). For both training and research purposes, it is important that construct and content validity are maximised. This means that if a driver is given training in hazard perception, variables in the scenarios reflect risk and not some other unforeseen factor. For scenarios to have good content validity, we would expect skilled practitioners to score higher than unskilled ones.

In top-of-the-range full-mission simulators we might expect the results on any given performance measure to be very similar to those that might be derived in a real system. So, for example, if a driver chooses a certain speed on a particular simulated road, we would expect them to choose a very similar speed on the real-world equivalent. The same would be hoped for in terms of lane placement, reaction times, gaps to other vehicles, overtaking choices, time taken to feel fatigued and so on (Bittner et al., 2002; Reimer et al., 2006). In practice, we find very few examples of systematic attempts to measure the objective validity of simulators. Instead, suppliers and practitioners are content with the concept of relative validity. For training purposes it is assumed that if appropriate behaviours are shown in a simulator, they will transfer, at least in part, to a real life situation. Similarly, researchers are satisfied that if two things are compared in a simulator (e.g., drugs, sleep regimes, information systems, vehicle settings), then if one promotes a better response than the other, then at least the direction of the effect would be expected to be shown in a real world equivalent trial too. For most research purposes, it could be argued that relative and face validity are the key factors. Face validity is important to encourage the participant to accept the nature of the investigation and engage in a natural way. Relative validity is important to enable generalisation of results to real world, equivalent situations. If the focus of the simulator is training or driver assessment, it could be argued that both performance validity and objective validity are of greater importance.

So, what *degree*, and how much *fidelity*, of simulation are necessary for effective and valid training of truck drivers? For maximum *face validity* of a truck simulator it would be necessary to specify the highest *degree* and *fidelity* available within a particular budget. But if the training has to be cost-effective when compared to traditional real world training, budgets will be constrained and a compromise might be needed in the views of *top-of-the-range* facilities. At present, there is little information available to enable perfect choices between expenditure on a particular motion system instead of on a particular visual system, or even sound and vibration system. Can we even say that motion is necessary for successful training?

The aviation literature provides a range of views. Some have suggested that because experienced pilots often rely on motion rather than instrument readings, motion becomes more important as experience level increases (Briggs and Wiener 1959, cited in Williges, Roscoe and Williges, 1973). Similarly, it might be argued that experienced truck drivers rely more on motion, sound and vibration rather than dashboard displays to judge the performance of the vehicle, whereas novice drivers might derive substantial benefit from systems that focus on instrument display. Thus, some training lessons appropriate for novice drivers might be conducted on part-task trainers (see Figure 9.10), but advanced skill-based lessons would require a motion component. If we decide motion is important, then fidelity must be addressed. Poor motion systems might not only have a negative transfer of training to real world situations, they will also lead to increased levels of simulation sickness (Reed, Diels and Parkes, 2007).

There are some very-high-fidelity, full-mission simulators in existence (see Figure 9.11), and whilst demonstrating that current and prospective technology can provide a dynamic and involving driving experience, such levels of sophistication come at a financial cost which may be unrealistic for the mass-training market. However, the prospective European Directive wording refers to *top-of-the-range*. Herein lies the difficulty. Not only may simulators that can be described as top-of-the-range, or high-end, possibly be over-specified, and out of the range of prospective users; the use of such terms implies that only systems that achieve some kind of high face and performance validity can have merit and value in training. There is, as yet, no distinction between full-mission and part-task simulators, nor acknowledgment

FIGURE 9.10   An example of a high-fidelity part-task trainer.

that realism (isomorphism with real road training) may not be necessary or even desirable in all circumstances.

Similar arguments might surround the fidelity of visual databases (road scenes). The simple view is that they need to be as realistic as possible. However, from a training perspective, that may not be correct. Certainly in terms of resolution, field of view, brightness, contrast and refresh rates, there seems value to having higher fidelity. However, it might be argued that the content of the visual scene itself does not have to be high fidelity (if that means close to photo-realistic representation of a real scene). There may be value in taking unimportant elements out of a visual scene,

FIGURE 9.11   The National Advanced Driving Simulator, Iowa, USA.

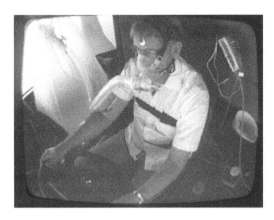

FIGURE 9.12   Experimental analysis of truck driver effort and stress at TRL.

allowing the driver to concentrate on elements salient to the training objective without distraction. Anyone involved in visual database development knows that there is a distinct law of diminishing returns (Figure 9.9, line B) to further expenditure beyond a certain point.

Williges, Roscoe and Williges (1973) proposed the notion of *essential realism*, relating not to what might be regarded as essential for improved face validity, but instead, essential to the particular training requirements under consideration. Indeed, face validity is the *bête noir* of training system specification, as it will always demand the highest feasible feature set (Figure 9.12). Instead, there are three important elements that should drive decisions on simulation provision within the training process:

The efficiency and acceptability of the learning in the simulator.
The transfer of the learning to the real world.
The retention of skills or knowledge learned.

Lee (2004) posed a number of interesting questions about simulator development and concluded that the pursuit of higher levels of fidelity in simulation may not be appropriate, or even desirable. The reasons being that increased fidelity can undermine scenario control, limit data collection, dilute training potential, and increase likelihood of causing simulator sickness. For example, if a simulation is able to provide a highly realistic and complex urban environment and a busy traffic situation, it may be a highly impressive demonstration of the state of the art of the simulation industry, but may force the driver to attend to elements peripheral to the current training objective. If the driver is supposed to focus on responding to a particular signal in the scene, a complex environment may present a number of competing signals and it would be difficult for the trainer to be certain which prompted the response by the driver. Taking extraneous noise from the scene (removing competing signals) would allow the particular behaviour or skill to be developed effectively and efficiently. The skill can then be later validated in a more complex and realistic environment, whether that is in a simulator or on the real road.

Reports are emerging that point to cost benefits of simulation training. Welles and Holdsworth (2000) reviewed features necessary to successful training in a range of commercial simulators and concluded that:

> … data to date, although sketchy, anecdotal or very preliminary, provides strong suggestion that driving simulators …. can reduce accidents, improve driver proficiency and safety awareness, and reduce fleet operations and maintenance costs.

They refer to hazard perception training with a particular police force leading to reductions in intersection accidents of around 74%, and overall accident reduction of around 24% in a 6-month period following training.

More recently, Dolan et al. (2003) presented evidence from a fuel management simulation study which tracked 40 drivers through a 2-hour training programme, and later for a 6-month follow up. Drivers were given specific training in the operational and tactical aspects of appropriate gear selection in a medium-fidelity simulator. Results indicated an average 2.8% improvement, with over 7% being indicated for those drivers with a poor pre-training record.

Such reports are encouraging, but do not take us far toward a minimum specification for systems that can provide a similar transfer of training benefit. There is a wealth of anecdotal experience that shows that road environments are difficult to model in simulators. Buildings and road signs have sharp edges, and road markings use white lines. At current typical screen resolutions, such scenes are prone to highlight aliasing and peripheral flicker, and text on signs will appear blurred. The number of polygons needing to be processed in a photo-realistic scene may also mean that simulators run at close to their processing limits and consequently display slow refresh rates.

One of the important skills in database creation is optimisation for the run-time environment to ensure that the scene can be processed efficiently (Figure 9.13). However, the starting point is usually taking realistic road layouts and scenes and

FIGURE 9.13   A road scene from the TRL DigiCar simulator 2010.

making adjustments to levels of detail and so on to make them usable, rather than coming from the other direction of taking what is known about cognitive processing, perceptions of speed and distance and producing an environment that includes salient information that allows training principles to be demonstrated, but which may be far removed from conventional views of realistic road scenes. Rizzo et al. (2002) have demonstrated an interesting concept with a database road scene created specifically to host experiments with cognitively impaired participants and their judgments of speed and distance. The experiment involved a large number of braking manoeuvres in a short space of time; and if conducted in a computer road scene of 'traditional' aspect would have been highly likely to promote high levels of simulator sickness. However, the specially constructed road environment used aspects of colour, tint, texture, and object obscuration coding to produce a driving environment which lacked face validity (apparent realism), but provided very high utility for the research team. Accurate data could be collected from a large number of participants with very minimal problems of simulator sickness.

The road scene below (Figure 9.14) is clearly an extreme example, but in being so, serves to illustrate the point that utility has been maximised by careful consideration of the particular requirements of the task under consideration.

Technological developments of simulators will continue as manufacturers seek to develop products for the marketplace and seek commercial advantage through performance improvement. However, improvements in computers and projectors, or motion systems, in themselves will not lead to a dramatic upswing in usage. A focus on *essential realism* is needed, and the main area for this is in the look and feel of the road scene databases. Such road scenes and scenarios must be developed that support the ability of the driver to interpret salient information without overloading the visual system with unnecessary information. The road scene must be a comfortable place for training to take place. Face validity, the *bête noir* of simulation development, must assume a lower priority for users, if real progress is to be made.

FIGURE 9.14 Example of a stylised road scene. (From Rizzo, M., Severson, J., Cremer, J., and Price, K., 200, An Abstract Virtual Environment Tool to Assess Decision-Making in Impaired Drivers. In J. D. Lee, M. Rizzo and D. V. McGehee (eds.), *Proceedings of the Second International Driving Symposium on Human Factors in Driver Assessment, Training, and Vehicle Design*, Iowa City: University of Iowa.)

Categorisation of training devices is a standard procedure in the aviation industry. Flight simulators are categorised into four bands, A, B, C and D according to the prescriptions of JAR-STD 1A.025 (Requirements for Flight Simulators (qualified) on or after April 1, 1998). The directive published by the Joint Aviation Authorities (JAAT) states that:

(a) Any flight simulator submitted for initial evaluation on or after 1 April 1998, will be evaluated against (applicable) JAR-STD 1A criteria for Qualification Levels A, B, C or D. (Recurrent) evaluations of a flight simulator will be based on the same version of JAR-STD 1A, which was applicable for its initial evaluation. An upgrade will be based on the currently applicable version of JAR-STD 1A.

According to this directive, flight simulators are assessed in those areas which are essential to completing the flight crew member training and checking process, including:

Longitudinal, lateral and directional handling qualities;
Performance on the surface and in the air;
Specific operations where applicable;
Flight deck configuration;
Functioning during normal, abnormal, emergency and, where applicable, non-normal operation;
Instructor station function and simulator control; and
Certain additional requirements depending on the Qualification Level and the installed equipment.

The JAAT directive specifies minimal technical requirements for simulators qualifying for JAA Level A, B, C and D, with A being the lowest and D being the highest level of technical complexity of the flight simulator. Certain requirements in the categorisation system must be supported with a statement of compliance (SOC) and, in some designated cases, an objective test.

A European Union collaborative research project (TRAIN-ALL) explored the possibility of transferring some of these principles to the realm of driving simulation. The project reviewed the published specifications of known training facilities and conducted a questionnaire survey of operators. It concluded that a banding system for the driving simulators could be developed in the style of the aviation industry, describing five levels, with level A being the lowest and level E being the highest level of complexity of the driving trainer (Lang and Parkes, 2009).

Five bands were identified rather than the four used in aviation because the technology clusters had different distinguishing characteristics. Most notably the aviation classification does not consider the very basic technology cluster of TRAIN-ALL Band A, but for driver training this could be an important component of the curriculum, and can be distinguished from other multi-media tools in that it has the important characteristic of demanding the student to take direct interactive control of a simulated vehicle in a traffic environment (similar to the concept of the original TRL simulator of the 1970s).

The suggested banding does not include all variables deemed relevant for a comprehensive benchmarking of driving simulators. Instead, it rather focuses on a few main variables that allow a rough classification of the systems under scrutiny in terms of their technical complexity. The suggested variables include:

Replication of vehicle features (e.g., controls, cab, sounds, kinaesthetic feedback);
Visual system (single versus multi-channel projection and field of view (FOV));
Motion rendition (none, basic, 6 and 8 degrees of freedom);
Interactivity/number of simulated road users (low, medium, high);
Sophistication of the simulated road environment, including road layouts and environmental conditions (low, medium, high);
Breadth of learning opportunities provided, e.g., complexity of training scenarios, changeability of underlying vehicle model and possibility to add driver assistance systems.

The sophistication of the simulated road environment was operationalised as the sum of simulated road environments (e.g., motorway or urban), the simulated road features (e.g., bridges or tunnels), simulated weather conditions (e.g., rain or night) and simulated lights (e.g., headlights or reversing lights).

The complexity of available training scenarios was calculated from the number of selected training goals on the manoeuvring level that a simulator could cover. The simulator questionnaire required respondents to specify which out of a total of 29 manoeuvring goals the training tool they described could address. These learning goals included the explicit interaction with other road users rather than just negotiation of static road environment features. The selection of items for the assessment is in line with the requirement from the EU driver training literature for a stronger focus of simulator training on higher-order skills.

Visual scanning.
Overtaking/parking.
Entering/leaving traffic.
Hazard perception.
Driving techniques in critical situations.
Reacting to other vehicles.
Reacting to vulnerable road users.
Negotiating junctions, intersections and roundabouts.

Typically, for low-band simulators most variables would assume low values, whereas training devices in the higher bands would be characterised by high values accordingly. As the suggested simulator banding includes a number of variables and at the same time, simulators can vary considerably in their capabilities, a perfect fit between banding criteria and simulator features is unlikely. The suggested banding approach allocates a simulator to the band it has the greatest overlap with on the basis of the existing information. It is also important to note that the suggested simulator banding is based on technical complexity of the driving simulators and does not imply a 'fit for purpose' judgment of the training tool itself. Low-band simulators may be very appropriate for achieving their intended training goals, for example, if the training goal is the familiarisation with car controls, the simulator will not be required to feature a sophisticated behaviour model of the simulated road users. Table 9.1 following, taken from Lang and Parkes (2009), gives a brief description of the general technical requirements associated with the five bands and the learning targets that are attainable.

There is no single dimension or attribute that allows a straightforward and non-controversial classification of driving simulators. Technology is advancing, and stakeholder expectations of systems are developing, not least because there is a general awareness of the developments being made in entertainment systems and in military-grade synthetic training. As shown, there are many sub-systems within even the most basic simulator-training device. Simple, 'level A' devices will employ a visual display, some form of manual input device, and a task presented within a traffic environment. As we move up the scale of face validity and complexity, motion, sound, and vibration are introduced, and the realism of the dynamic driving task increases. The fact that different systems may have strengths in one area but weaknesses in others, compared to systems of roughly similar price and training aim, means that a classification which takes a technology focus struggles to provide distinct immutable classes.

The assignment of simulators to the bands A through E is not intended by the authors to be definitive, nor should it be viewed as having importance beyond that of providing a working example of how such a system might operate if full information were available for each system under consideration.

In order that purchasers of simulator-based training can have confidence that they will receive value, they will need to know the curriculum is to be delivered through the appropriate training medium. That implies a clear knowledge that the particular system has been accredited as appropriate. Thus, such a banding system might be an important preliminary step towards an acceptable classification system.

Future consideration will be needed of how to accommodate forthcoming developments in advanced vehicle displays and controls (head-up displays, voice-activated controls etc.), and also systems such as navigation, adaptive cruise control, collision avoidance, vision enhancement and so on. It may not be the case that interaction with advanced systems can only be trained on the most sophisticated levels of simulator.

At present simulators are focused on providing the opportunity for training in skill-level operational control and for tactical decision making in terms of responses to potential hazards in the road environment. They have not been developed with strategic level decision making in mind (route choice for example), yet these aspects will form an increasing component of future training programmes.

It is also likely that some forms of training, for example of emergency service drivers, will require increasing emphasis on direct participation by the trainer in the form of manipulation and control of other road users (drivers or pedestrians) in the road scene. At present there are very few systems that allow the trainer direct control of other vehicles in the road scene via a separate instructor station.

The training experience is a function of the characteristics of the trainee, the trainer, the training delivery system, and the dynamics of the interaction between them. As such, although the content of the trainer function is likely to change in the future

TABLE 10.1

Potential Classification Bands for Training Simulators

Qualification Level	General Technical Requirements	Learning Targets
A	*The lowest level of driving simulator technical complexity*	Suitable for:
	The driving simulator enables the user to navigate the ego-vehicle through a populated road environment displayed on a single channel screen. Rear and side mirror views may not be provided.	Awareness raising, visual familiarisation with road environments, or simple entertainment
	The movements (vertical and lateral) of the ego-vehicle are controlled by the use of a mock steering wheel and pedals or by joy-stick. Kinesthetic feedback for driving controls is not provided.	(May have value in promoting life goals and strategic issues)
	The driving simulation does not include realistic gearshifts, a vehicle cabin or a motion system.	
	Changes of the underlying vehicle model are not possible or very limited.	
	The interactivity of simulated road users is low.	
	Low number of simulated road environments and environmental conditions.	
B	As for Level A plus:	As for Level A plus:
	Provision of car controls, including pedals, gearshift, steering wheel.	Familiarisation with vehicle controls and procedures possible. Compliance with some rules of the road.
	Provision of an artificial vehicle cab.	
	The road environment is displayed on a single visual channel.	
	Side or rear mirror views may not be provided.	
	No motion system provided.	
	Limited number and interactivity of simulated other road users.	
	Limited number and realism of simulated road environments and environmental conditions.	
C	As for Level B plus:	As for Level B plus:
	Realistic feel of car controls (e.g., pedal or steering wheel resistance).	Training of simple manoeuvring tasks in small number of road environments possible; and some tactical decision making in simple traffic.
	A motion system may be provided.	
	Wider FOV through multi-channel projection often provided.	
	Greater number and realism of simulated road user behaviour.	
	Training of more complex driving scenarios possible.	

(Continued)

TABLE 9.1 (CONTINUED)
Potential Classification Bands for Training Simulators

Qualification Level	General Technical Requirements	Learning Targets
D	*The second highest level of driving simulator complexity*	
	As for Level C plus:	As for Level C plus:
	Provision of realistic vehicle cab.	Training of complex
	Multi-channel visual system with provision of rear and side mirror views.	manoeuvring tasks including interaction with other road
	6 degrees of freedom motion system provided.	users, hazard perception, and
	Larger number and realism of simulated road environments and environmental conditions.	eco-driving.
	Behaviour of other road users can be influenced.	
	High degree of interactivity with simulated road users provided.	
	Change of underlying vehicle model possible.	
	Addition of advanced driver assistance systems possible.	
E	*The highest level of driving simulator complexity*	
	As for Level D plus:	As for Level D plus:
	6 degrees of freedom plus extended x and y motion system (rails) provided.	Wider range of complex manoeuvring tasks recreated
	Training with highly complex training scenarios with high level of interactivity between road users.	adequately due to availability of more comprehensive motion rendering.

as synthetic training becomes more widely adopted, the importance of the trainer as facilitator, mentor and confidant of the trainee will continue, and be particularly important for those trainees who might not be at ease with the technologies involved.

Movement towards certification and accreditation of driving simulation systems is needed but will take a long time to reach consensus. Debates will continue about how best to consider simulator fidelity and validity and the establishment of agreed criteria for simulator categorisation will emerge only slowly. In the meantime, simulators will continue to develop, expensive and ever grander motion systems will be commissioned, better-resolution, computer-generated images will be projected and greater attention will be paid to the realistic feedback of control forces. It is less clear that similar attention will be paid to resolving how best to fit a research simulator to a particular research issue, or how best to integrate synthetic training into the general training programme.

This chapter started with reference to the legacy of aviation trainers in the development of driving simulators. It is possible that we have been guilty of learning some of the wrong lessons. Instead of being drawn inexorably towards the high-end, full-mission simulators that undoubtedly have their place in flight training, we should also have recognised that there is a place for high-fidelity, part-task simulators for both training and research purposes and a focus on marrying the technology to the primacy of the research or training requirement. A movement away from validity and back towards essential realism is needed.

REFERENCES

Bittner, A. S., Simsek, O. Levison, W. and Campbell, J. (2002). On-road versus simulator data in driver model development. *Transportation Research Record*. Vol. 188, 38–34.

Briggs, G. E. and Wiener, E. L. (1959). Fidelity of simulation: Time sharing requirements and control loading as factors in transfer of training. Orlando: *NAUTRADEVCEN*. 508–5.

Carsten, O. and Jamson, H. (2011). Driving simulators as research tools in traffic psychology. In B. Porter (ed.) *Handbook of Traffic Psychology*. San Diego: Academic Press, pp. 86–96.

Dolan, D. M., Rupp, D. A., Allen, J. R., Strayer, D. L. and Drews, F. A. (2003). Simulator training improves driver efficiency: Transfer from simulator to real world. *Proceedings of Second International Driving Symposium on Human Factors in Driver Assessment, Training and Vehicle Design*, Park City, Utah, US.

DfT (2003). *The Safe and Fuel Efficient Driving (SAFED) Standard. Good Practice Guide 2100*. London: Department for Transport.

EU Commission (2003). Directive 2003/59/EC of the European Parliament on the initial qualification and periodic training of drivers of certain road vehicles for the carriage of goods or passengers. *Official Journal* L226 P.0004–0017.

FWHA (1999). Research design: Validation of simulation technology in the training testing, and licensing of tractor-trailer drivers. Report no. FWHA-MC-990060. US Department of Transportation.

Godley, S., Triggs, T. and Fildes, B. (2002). Driving simulator validation for speech research. *Accident Analysis and Prevention*. Vol. 35 (5), 589–600.

JAAT (1998). Joint aviation requirements: Aeroplane flight simulators. JAR-STD IA.025. Joint Aviation Authorities Committee. Cheltenham: Westward Digital Limited.

Kaptein, N. A., Theeuwes, J. and Van der Horst, R. (2007). Driving simulator validity: Some considerations. *Transport Research Record*, Vol. 1550, 30–36.

Kalawski, R. (2000).Validity of presence as a reliable human performance metric in immersive environments. *3rd. Workshop on Presence*, the Netherlands. http://www.temple.edu/ispr/prev-conferences/www.preence-research.org/kalawski.pdf, accessed: January 20, 2012.

Lang, B. and Parkes, A. M. (2009). Benchmarking and classification of training simulators in driver training. *Proceedings of Second Technology Based Training for Drivers (TTD) Conference*, January 21–23, 2009. Bonn, Germany. Deutscher Verkehrssicherheitsrat. (CD Rom).

Lee (2004). How low can you go? *Proceedings of Human Factors and Ergonomics Society Conference*, New Orleans. (CD).

Luke, T., Parkes, A. M. and Walker, R. (2006). The effect of visual properties of the simulated environment on driver behaviour and simulator sickness. *Proceedings Driver Simulation Conference (DSC Europe) 2006*, pp. 253–262.

Parkes, A. M., (2003). Truck driver training using simulation in England. In J. D. Lee, M. Rizzo and V. McGehee (eds.), *Proceedings of the Second International Driving Symposium on Human Factors in Driver Assessment, Training, and Vehicle Design*. Iowa City: University of Iowa. pp. 59–63.

Parkes, A. M. (2005). TruckSim: Cost-benefit of simulation for truck driver training. *Proceedings of Technology based Training for Drivers Conference*, November 17–18, 2005. Dresden, Germany: Deutscher Verkehrssicherheitsrat. (CD Rom), PATRS508105.

Parkes, A. M. and Reed, N. (2005). TRUCKSIM: Preliminary results from cohort study in England. *Proceedings of HUMANIST Conference on Application of New Technologies to Driver Training*, Brno, Czech Republic, January.

Parkes, A. M. and Reed, N. (2005). Fuel efficiency training in a full-mission simulator. *Behavioural Research in Road Safety 2005. Fifteenth Seminar*. London: Department for Transport, pp. 135–146.

Parkes, A. M. and Reed, N. (2006). Transfer of fuel efficient driving technique from the simulator to the road: Steps towards a cost-benefit model for synthetic training. In D. de Waard, K. A. Brookhuis and A. Toffetti (eds.), *Developments in Human Factors in Transportation, Design and Evaluation*. Maastricht, the Netherlands: Shaker Publishing. 163–176. (TRL PATRS508305).

Reed, N., Basacik, D., Chattington, M., and Parkes, A. M. (2009). A methodology for the investigation of distraction by advertising using a driving simulator. *DSC Europe 2009*, February 3–5, Monaco.

Reed, N., Diels, C. and Parkes, A. M. (2007). Simulator sickness management: Enhanced familiarisation and screening processes. *The First International Symposium on Visually Induced Motion Sickness, Fatigue, and Photosensitive Epileptic Seizures* (VIMS2007), Hong Kong.

Reed, N. Diels, C. and Parkes, A. M. (2008). Validation of participant screening processes for simulator sickness management. *DSC Asia*, Seoul, Korea 2008 (CD Rom).

Reed, N. and Parkes, A. M. (2005). Correlates of simulator sickness in a truck driver training programme and the development of an effective screening process. In de Waard, D., Hockey, B., Nickel, P. (eds.) *Human Factors Issues in Complex System Performance*. Maastricht, the Netherlands: Shaker Publishing.

Reimer, B., D'Ambrosio, L., Coughlin, J., Kafrissen, M. E. and Bierderman, J. (2006). Using self reported data to assess the validity of driving simulator data. *Behaviour Research Methods*. Vol. 38, No. 2, 314–324.

Rizzo, M., Severson, J., Cremer, J., and Price, K. (2002). An abstract virtual environment tool to assess decision-making in impaired drivers. In J. D. Lee, M. Rizzo and D. V. McGehee (eds.), *Proceedings of the Second International Driving Symposium on Human Factors in Driver Assessment, Training, and Vehicle Design*. Iowa City: University of Iowa.

Welles, R. T. and Holdsworth, M. (2000).Tactical driver training using simulation. I/ITSEC 200 Conference. November 30, 2000. Orlando, Florida, US.

Williges, B. H., Roscoe, S. N. and Williges, R. C. (1973). Synthetic flight training revisited. *Human Factors*, 15, 543–560.

Witmer, B. G.. and Singer, M. J. (1998). Measuring presence in virtual environments: A presence questionnaire. *Presence*. Vol. 7, 225–240.

10 Human–Machine Interaction (HMI) in the Time of Electric Vehicles

Nikolaos Gkikas
Autonomics, UK

CONTENTS

Over a century ago in Mannheim, Karl Benz installed a combustion engine to his infamous tricycle, which is widely accepted as the first automobile. Whether that (see Figure 10.1) was truly the case or not will always remain open to argument; what however can hardly be denied is the rapid development of the automobile to the ubiquitous transport mode and instrument for recreation it is today. In highly motorized countries such as the US, Japan and in Europe, it is difficult to imagine life without road vehicles. In other developed countries, more often than not, a car is the most desirable purchase in a person's life, to the point of sometimes competing with the purchase of a house.

10.1 BACK TO THE FUTURE …

In-between then and today, vehicle development has been relentless, and looking at any of the early examples of the automobile, it is very difficult to identify automotive parts that withstood those 120–130 years. Even the most basic of parts by today's

FIGURE 10.1   The start of motoring?

standards, such as a steering wheel or pedals were quite uncommon then. One hundred years ago, the steering wheel was a novelty. Electric cars however, were not ...

Far from it, in the early 1900s electric vehicles were the norm rather than the exception: they were more efficient, far less noisy and dirty than their combustion-powered siblings and did not suffer any relative range inefficiencies. It was only in the second decade of the century, and through the momentum created by the Great War that internal combustion engines (ICE) witnessed huge development for the needs of the first major motorized conflict. In addition, ICEs could be employed in a variety of transport applications—from featherlight fighter planes to the gargantuan battleships of the time. Electric vehicle production peaked in 1912 (About, 2012). From then on, the tide turned against the electric motor. The ICE witnessed constant development, which continues to the present day. Electric vehicles (EV) on the other hand lost any prior competitive advantage and became virtually extinct.

The second half of the 20th century witnessed the comprehensive standardisation of vehicle characteristics and their fundamental ergonomics. That effectively happened both through the formal route of engineering (e.g., SAE), national (e.g., DIN, ANSI, BSI) and international standards (ISO), but also through the informal request of the society for product reliability, and cost-reducing efficiency. From an ergonomics/human factors engineering perspective, that trend led to reduced human error and accidents in the production line as well as improved familiarisation and reduced training demands for drivers changing from one vehicle to another. It also meant that radical engineering ideas would have a very hard time—if any chance at all—to make it to the showroom.

Some alternative ideas however are strong enough to receive the support of stakeholders (manufacturers, policy-makers and/or the public) and endure the long filtering and cost/benefit weighing processes in place. Among those, the return of the electric vehicle (EV) has received significant support from some manufacturers, governments and part of the public, predominantly due to its green credentials and the holy grail of emission-free motoring it carries. When these lines were typed, already seven mass-produced electric vehicles were available in the markets of Japan, US and Europe.

10.2   ELECTRIC, HYBRID, PLUG-IN HYBRID, FUEL CELLS AND ...

Externally, it is difficult to distinguish an EV among other vehicles in a car park. Even when it starts moving, other than the absence of a tailpipe and the low levels of noise, there is very little to pick in between. So what is an EV and how different is it from other vehicles? Figure 10.2 illustrates the fundamental differences between an EV and an ICE powered vehicle. In an ICE vehicle, the energy source (fuel) is stored in a tank, which is hydraulically (pipes) connected to the ICE. Fuel is pumped to the ICE, which in turn transforms the chemical energy of the fuel into kinetic energy. Due to the revving characteristics of the ICE (see Figure 10.4 in the following section), a gearbox is necessary before the kinetic energy reaches the driven wheels. In an EV, energy is stored in a charged battery, which is electrically linked to the motor. The motor—through an intermediate inverter, in the case of an AC motor—transforms the electrical current from the battery to kinetic energy transmitted directly to the driven wheels.

The above applies to 'pure' electric vehicles. However, already from the mid-1990s, mass-produced hybrid gasoline–electric vehicles made their appearance

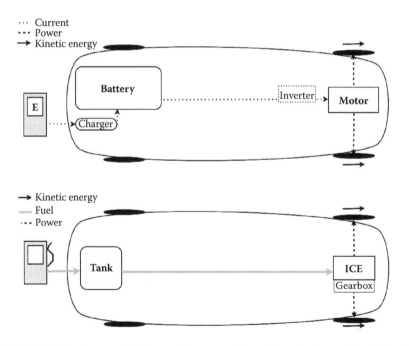

FIGURE 10.2   Fundamental differences between an internal combustion engine (ICE) vehicle and an electric vehicle (EV).

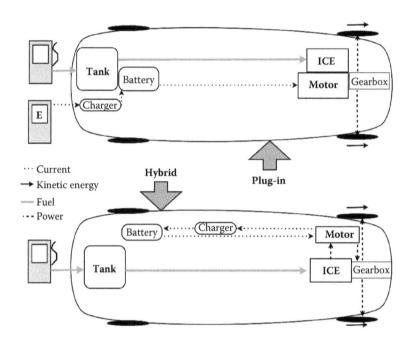

FIGURE 10.3   Typical hybrid and plug-in hybrid powertrain layout.

(see Figure 10.3). The idea was to employ a small electric motor alongside the ICE in order to reduce fuel consumption. The motor also acts as a generator, charging the battery under braking. Thus, energy otherwise wasted as heat (brake disks or drums), is now partly recovered, transformed to current and stored in the battery. In a hybrid, the battery is usually many times smaller than in a pure EV. Accordingly, the effects of electric power are limited. For that reason, plug-in hybrids have been developed. At the time this manuscript was written, no plug-in hybrids were available in the mass market; however, a number of them were expected within a year. What makes them special is the fact that they bring hybrids closer to pure EVs. Battery size is significantly increased, and instead of relying on the—limited—recovered energy while decelerating, a charger is in place and the battery can be charged in the same way as a pure EV. Thus, short-distance trips can be completed on electric power only, while the ICE takes over when the battery is depleted.

Using components similar to those in a plug-in hybrid—albeit a different arrangement which brings it closer to a 'pure' EV, instead of driving the wheels when the battery is depleted, range-extender EVs employ an ICE only as generator to charge the battery. This is an even more niche, however promising, powertrain solution (again at the time of this manuscript). The obvious benefit: optimisation of the ICE to work only as generator results in improved efficiency (engine always revs at its optimum range). Another rare powertrain solution is the hydrogen/fuel cell electric vehicle; the addition of fuel cells to an EV platform and the replacement of the battery with a hydrogen tank turns a vehicle into a water-producing factory (for details, see Appleby, 1988). Of course, there are practical obstacles to the generalisation

TABLE 10.1

Basic Characteristics of Electric and Hybrid Electric Vehicles

	Electric			Hybrid	
Power source	Battery	Fuel cell	Battery + generator (fuel)	Gasoline/diesel + battery	Gasoline/diesel (+ battery)
Powertrain	Electric motor	Electric motor	Electric motor	ICE + electric motor	ICE (+ electric motor)
Efficiency	High	High	High–medium	Medium	Medium
Energy capacity at given volume	Low	Medium	Medium–high	High	High
Conventional name	Pure EV	Hydrogen vehicle	Range extender EV	Plug-in hybrid	Hybrid

of this exciting technology and only a single example of such vehicle is currently available in the market.

Table 10.1 presents a summary of the key characteristics of electric and hybrid electric vehicles. The classification should not be seen as absolute; we already witnessed the thin line between plug-in hybrid and range-extender EVs. Let us not forget that hybrids themselves partly are electric vehicles. For the purpose of understanding the relative merits of each solution and its impact on driving ergonomics and human–machine interaction (HMI) however, it is useful and should be sufficiently valid. Starting from the right column, we have hybrids and plug-in hybrids. Both of these types use an electric motor alongside the ICE, as illustrated on Figure 10.3 above. Both hybrids tend to have lower emissions than their ICE counterparts, however in reality this is very much **dependent on driving style and behaviour**. On the other hand, by retaining the fuel tank of the equivalent ICE vehicle, they can store high amounts of (chemical) energy. Therefore, they often match or exceed an ICE vehicle's range. The plug-in type may exhibit further increased efficiency depending on **how use determines work ratios between ICE and electric motor**. Next to the plug-in hybrids, range-extender EVs appear more efficient, thanks to the improved efficiency of their generator compared to an ICE used to drive wheels. This is because, unlike an ICE driving wheels, a generator can spin constantly at its optimum frequency to charge the battery. In addition, range-extenders often (but not always) have a larger battery. This allows them to work as emission-free EVs for longer periods of time. Again, the above may often not be realised, depending on **how a vehicle is driven.** The fuel cell EV, like the range-extender, uses a generator to produce electricity and power the motor. That is where the similarities end however, as the fuel cells do not 'burn fuel'. They allow hydrogen to react with oxygen from the air and produce water. Electrical current is also released during the reaction and that is what goes to the motor that drives the wheels. Thus, a decent-sized hydrogen tank is required; however, due to the chemical properties of hydrogen (gas/low density even in liquid form) and current efficiency of fuel cells, a same-size hydrogen tank provides shorter range than a fuel tank for the equivalent ICE. The main hindrance to fuel cell generalisation is **availability and accessibility** to hydrogen in a usable form. By contrast, electric power is virtually everywhere; 'pure' EVs can draw power from most sockets in modern houses, shops, offices, workshops and car parks. In addition, like all EVs, they enjoy the super-efficiency and **driver-friendly properties** (see next section) of the electric motor. Thus, they consume minimum energy for a given distance. Their weakness lies in the energy capacity of batteries, which can store many times less energy in the same volume as a fuel tank. Ratios are not fixed, vary from fuel to fuel and depend on many parameters, however it is fair to say that with current lithium-ion technology, it would take a few decades of same-size battery packs to store the equivalent energy stored in a gasoline tank.

10.3 THE PROPERTIES OF THE ELECTRIC MOTOR

The electric motor is arguably the most influential difference between EV and ICE vehicles. That is because its revving characteristics are very different, almost opposite to those typically expected from an ICE. These differences are apparent in all three key parameters of motor efficiency: energy efficiency, torque and power output characteristics.

In terms of energy efficiency, there is a clear and comprehensive advantage of the electric motor over the ICE; a gasoline engine typically has less than 30% overall efficiency (Motlagh et al. 2008), that is, less than 30% of the chemical energy of the fuel is transformed into kinetic energy that moves the vehicle. On the other hand, even a standard electric motor is expected to display an efficiency over 90% (Keljik, 2009). The difference in efficiency is clear and relatively easy to comprehend. The comparison on torque and power output requires more explanation, as the use of instant or average numbers can be misleading without a good idea of performance over the whole rev range of the engines. For that purpose, Figure 10.4 presents the torque and power graphs for two comparable—in terms of maximum power output—powertrains; an electric motor and a modern high-efficiency gasoline engine.

The electric motor starts from zero revs with maximum torque. Thus, the motor may increase its rhythm quite easily for the first few thousand revs. In parallel, the

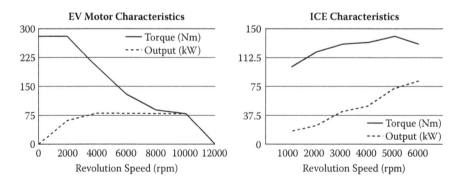

FIGURE 10.4 The contrast between EV and ICE powertrain characteristics.

power output increases linearly up to a certain peak in the first part of the rev-range. Approximately at the same point, torque starts to drop gradually. Power however forms a plateau, with the engine retaining the top output for its remaining range. It is only due to some phenomena, outside the remit of this book, that when torque is sufficiently dropped, power follows suit as the motor reaches its nominal limit. The ICE, on the other hand, has its lowest torque at idle (usually 700–1,000 rpm). As a consequence, the rhythm of revolution does not increase as quickly as at higher rpm and certainly not as effortlessly as in the case of the electric motor. Power increases in accordance with torque, although not as linearly or steadily as in the electric motor. Nevertheless, power keeps increasing even after torque reaches its maximum and starts declining. It is not uncommon in ICE for maximum power to be delivered right at the rev limiter.

Aside from the different patterns in torque characteristics and power output, it is worth noticing the respective numbers associated with the trends presented in Figure 10.4. First, the ICE is typically limited to around 6,000 rpm (for gasoline engines; limit is usually lower for diesel engines). The electric motor retains its peak power output all the way to 10,000–11,000 rpm. The ICE delivers the top 50% of its power between 4,500–6,000 rpm, while the electric motor delivers 80% of its peak power already from 2,000 rpm and 100% from 4,000 to 11,000 rpm. In Figure 10.4, EV torque appears to drop dramatically between 2,000–6,000 rpm; however, even at 6,000 rpm it matches the torque of the ICE at the same rpm (130 Nm). It is only when it exceeds the rpm range of an ICE (6,000–11,000 rpm) that it falls below the typical toque range of an ICE. The latter has obvious human factors implications both to the engineering of the vehicle as well as to how a driver experiences the vehicle.

10.4   THE EV DRIVING TASK

At this point, it would only be reasonable for one to wonder why in a chapter about ergonomics and HMI it was necessary for two sections to be dedicated to the engineering background of EVs. The answer to that question is that every detail presented in the sections preceding has a direct or indirect impact on how a driver experiences the vehicle, what specific information and control requirements are set, and how vehicle engineering needs to adapt in order to address those. In a nutshell: the ergonomics/human factors/HMI of such vehicles.

10.4.1   How EVs Change the Driving Task

It is now over 40 years since the extensive task analysis of the driving task and the subsequent control-model by Allen, Lunenfeld and Alexander (1971). Their model suggested a three-level architecture of the driving task: at the bottom, the maneuvering level consisted of the basic control tasks: from steering inputs and gear shifting to operating the wipers and managing lead headway in traffic. Above that, the tactical level consisted of tasks that require some conscious decision making, often in response to the changing traffic environment: deciding which route to take home, taking a shortcut or not, adapting speed when weather conditions change or going through a roadworks area. At the top, the strategic level of driving includes highly demanding cognitive tasks, learned behaviours, attitudes, and even beliefs that precipitate the relationship with the vehicle, other road users and the road environment: from the problem-solving mechanisms of plotting a route in a totally unfamiliar area, to general attitudes towards speeding and risk taking, vehicle preferences, driving style and preferences, presumptions about other drivers, riders and pedestrians, and so forth.

Within those 40 years, this model has been developed and adapted to explain specific facets of driving (e.g., Michon, 1985; Summala, 1996; Lee and Strayer, 2004); however, the core of those that followed remains true to the original and the three-level control model has been widely accepted as a valid description of the driving task (Lee, 2005). As vehicle and road engineering, driver training and expectations (from drivers) evolve, the details for each level are subject to change. The introduction of in-vehicle information systems (IVIS) and advanced driver assistance systems (ADAS) has already affected the driving task, with obvious changes taking place in the maneuvering level (Gkikas, in press). The particular characteristics of EVs presented in the previous sections have their own impact on the details of the driving task.

Figure 10.5 presents the basic maneuvering-control level of the contemporary driving task. Tasks marked with an asterisk are affected by the move to EV and tasks marked with double asterisk are either introduced by or significantly altered in EV. The approach to an EV, entry and setting the driving position is hard to distinguish from the same procedure in any other modern vehicle. It is only when configuring indirect vision systems (mirrors or cameras) where some additional attention should be paid. With the absence of the ICE cues of vibration and noise when stationary, it is important that a countermeasure with appropriate human factors/ergonomics specifications is engineered (e.g., some type of auditory warning or a moving object recognition system); in addition, some skill development in anticipatory control is beneficial—at least until EVs become common and other road users become familiar with their characteristics.

Immediately before or after setting up the driving posture, the driver in an ICE vehicle switches electrics and then the ignition on. In an EV, those two actions effectively become a single action accomplished with the start switch/button. When the electrics come on, the motor is ready to respond when the drive mode is selected (e.g., 'Drive', 'Reverse', 'Forward') and the brake pedal is released or the throttle pedal is depressed. Although desirable, this readiness of the vehicle requires some familiarisation, especially for drivers with prolonged experience of ICE. Otherwise, misunderstanding the 'vehicle status' is quite possible, confusion and dissatisfaction almost certain and safety implications (e.g., pulling away inadvertently) possible.

To start any trip, the ICE driver has to select first gear (in manual transmission), 'D' (in automatic transmission) or reverse gear, release gently the clutch or brake pedal and depress gently the accelerator. In an EV, the process is usually close to an ICE with automatic transmission, although there are some subtle differences. As we witnessed in the previous section, EV motors have a much longer revving range than ICEs and much higher torque available from 0 revs. As a result, EV may not need a gearbox intermediately between the motor and the transmission shafts to the

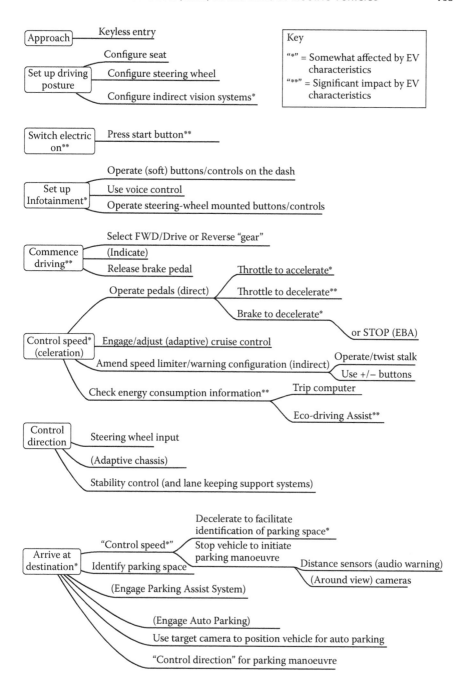

FIGURE 10.5 EV driving task at manoeuvring-control level.

wheels—no need for lower gears due to high torque and no need for higher gears due to long revving range. The direct consequence is that unless that—beneficial in every other way—toque is somehow 'damped', then the sudden pull of the motor will be difficult for most drivers to control. Scenes such as pulling out of a tight parking space in the city without causing damage would be impossible for some drivers. For that reason, EV special accelerator, pedal tune and motor management is required. For any EV to be successful in the contemporary and future market, it would have to provide ICE-level controllability for acceleration from stop.

The same applies while driving en-route. Special pedal tune is required to provide 'natural' control of acceleration. Furthermore, EVs use regenerative braking to charge the battery when acceleration is not required. That function reverses the motor function into a generator that charges the battery. From an ergonomics point of view, the obvious effect is that the accelerator pedal effectively becomes a reversed brake pedal, which is applied when released. Contemporary technology in energy recovery systems is still limited in braking capability compared to traditional hydro-mechanical or electromechanical (brake-by-wire) systems. Therefore, the primary braking system is, as in the case of ICE vehicles, controlled through a brake pedal. Nonetheless, how regenerative braking is controlled, by the accelerator or a combination of accelerator and brake pedal, and the tuning of the system is a critical human factors engineering challenge. The issue is discussed further towards the end of the chapter.

With ecological concerns over road transport and economic concerns over the price of fossil fuel, some form of eco-driving advice/display is available in most modern vehicles, starting from the simple trip computer that displays mean and instant consumption figures to graphic displays and gear-shift indicators that facilitate efficient use of the ICE. Considering the reasons behind the purchase of an EV, it is easy to imagine how important eco-driving driving support is. In addition, with the stored energy limited by the capacity of the battery, efficient use of energy becomes a pragmatic requirement for the everyday use of the vehicle. For this reason, EVs need to take eco-driving support to another level. Energy consumption information needs to be displayed in a direct and efficient manner to the driver. Therefore, modern vehicles tend to display energy consumption with a gauge in the instrument cluster—where one would normally find the tachometer (or the speedometer) in an ICE vehicle (Figure 10.6). Optimisation of eco-driving displays is not enough in EVs; control of energy consumption is required and this translates not only in the tune of the accelerator pedal, as already has been mentioned, but also to the provision of an active-feedback pedal. That feedback needs to be tactile due to the nature and position of the control in relation to the driver; it could either take the form of vibration or variable resistance. The successful specification of each of such systems carries its own ergonomic challenges (Young, Birrell and Stanton, 2011), and as mentioned above, integration with throttle management on one hand and regenerative braking on the other can be decisive to the success or failure of an EV.

Another critical area of modern vehicle HMI is interaction with IVIS (see also Chapter 4 in this book). Even more so in the case of an EV, navigational information should not be limited to the quickest route to destination, the shortest distance or the one with the minimum amount of traffic congestion. Such information, although

FIGURE 10.6 The instrument cluster in an EV.

relevant, is of little benefit to the EV driver without complementary information on energy consumption characteristics of the route (elevation profile, traffic flow and opportunities for efficient energy recovery), and availability of (quick) charging stations. Once more, a holistic approach where the aforementioned requirements are integrated to the specification of vehicle systems, and drivers have achieved some basic familiarity with the particulars of EVs is necessary.

Finally, after all the tasks on Figure 10.5 have been performed successfully for as many times as required during a trip, the EV arrives at its destination. For the purpose of identifying a parking space, a combination of IVIS information (e.g., car park coordinates), ADAS (park assist; if available) and direct visual-motor control of the vehicle by the driver is required in the same way as it is in an ICE vehicle. The particulars of the EV descend from the same acceleration and braking characteristics that were described previously; initial motor torque control (and how user-friendly accelerator-pedal tune is), the starting process and awareness of drive mode and 'what the vehicle is going to do next', and properties of regenerative braking.

10.4.2 How the Electric Motor Changes Vehicle HMI

So far, we have witnessed the fundamental changes in vehicle engineering and their impact on the basic maneuvering-control level of the driving task. In the latter section, some comments were made about how in line with the driving task changes in the human factors specification of the vehicle are required. In this section, we visit the characteristics of control and display—fundamental HMI—for EV.

10.4.2.1 Information Display

We already witnessed in the previous section some scenes where EV driver information needs are different to those of an ICE vehicle driver. Beyond the basic maneuvering level, information plays an important role to the tactical and the strategic level of driving (Michon, 1993). The predominant differences in an EV descend from the prioritisation of energy consumption information, both at a strategic and a tactical level; early adopters of EVs are either driven by ecological or economic criteria—or both. Otherwise the purchase of an EV makes little sense, although, at the policy level, the criteria are explicitly ecological (European Commission, 2010). To satisfy both ecological and economic criteria, real-world energy consumption in an EV is key. Going one step down to the tactical level of driving, the reality of the energy characteristics of the motor and the battery put the reality of living with an EV to the test. Here it is not only about being efficient for the environment or for the accountant's sake. It is about satisfying the daily needs of getting to the destination stress-free and in the most enjoyable way possible. So, what does the driver need to fulfill the tactical requirements of EV motoring?

It was mentioned previously that navigational information needs to include energy demand prediction and charging facilities, as well as quickest and shortest routes. Many electric vehicles have included instantaneous energy consumption information by replacing the tachometer of an ICE vehicle with an energy consumption gauge (Figure 10.6). Nevertheless, instant information, although useful as an eco-driving advice (Birrell, in press), is of limited use when daily routes to and from destinations are planned. Economically, an EV should be used regularly in order to justify the premium over an ICE paid at point of sale. To achieve the contrasting requirements for extensive use and prudence with energy, a combination of information must be available to the driver in an appropriate format for quick comprehension with minimum distraction. With regards to the appropriate format, well-documented advice can be found in many other reference books (e.g., see Sanders and McCormick, 1993 for fundamentals of information display design). Regarding the type and contents of information, Figure 10.7 presents the four information elements required as

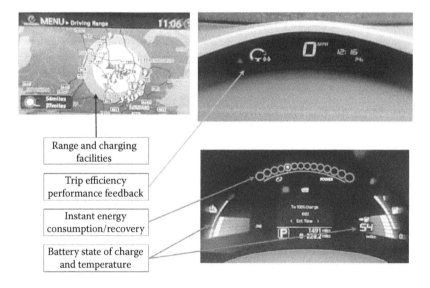

FIGURE 10.7 Combination of information-display units in a modern EV.

instantiated in a contemporary EV. Those are: estimated range and charging points (both visualised on the area map), instant flow of energy, battery information, and some form of longitudinal efficiency feedback based on the trip distance already covered. The importance of the first element has already been underlined many times in this chapter. The energy or power gauge serves two purposes. Information about battery state of charge corresponds to the fuel gauge in an ICE, and it is easy to understand its significance. The need for information on battery temperature on the other hand is not as straightforward; changes in temperature influence battery performance and subsequently the amount of energy available at any given time. Thus, awareness of any such deviation is useful to the driver as prediction of impeding reduction in effective range. The good news is that this will hardly ever be the case, and only in extreme climates, because every modern EV should be equipped with some form of heat management system for the battery.

10.4.2.2   Vehicle Control

Vehicle control consists of longitudinal control, which is typically realised through the operation of pedals and the gear level, and lateral control, which is realised through the operation of the steering wheel. With regards to the former, the particular characteristics of the electric motor require appropriate human factors engineering input to pedal tuning and power management. They also provide an excellent opportunity to facilitate efficiency without compromising the driving experience in an EV.

The high torque characteristics of the electric motor allow for stronger acceleration from stop than equivalent ICEs and at many times lower cost in energy. In fact, EVs change the traditional characteristics of eco-driving with an ICE: retain highest gear within the efficient rev range of the engine, avoid using the brakes and stopping, accelerate and decelerate smoothly. With the exception of the latter (being smooth with vehicle controls is always desirable), EVs somewhat invert the first two. Being in the highest gear in a modern ICE vehicle (long gears for economy) in practice suggests a speed above 70 mph/115 kph. Due to the nearly double revving range compared to an ICE, an electric motor is directly connected to the driven wheels and subsequently achieves such speeds at higher revs. Higher revs however, come with lower torque for the same power output. At the same time, aero drag becomes the main source of resistance to vehicle movement. Therefore, the EV consumes much more energy to retain such speed. By contrast, in a stop-and-go traffic scene, where in an ICE vehicle most of the energy is wasted through braking, the EV can harvest significant part of the energy used, through regenerative braking. Ignoring the energy spent to overcome the rolling resistance of the tyres, very little energy is otherwise wasted. In practice, most drivers will be surprised by how much energy they use doing motorway/highway speeds and how little energy they use in stop-and-go traffic.

Another effect of the direct link from motor to driven wheels is the much simpler 'gear' selector in an EV (see Figure 10.8). Instead of a gear selector, EVs require a simple drive selector for the forward and backward movement of the vehicle and a 'park' setting for the duration the vehicle is parked. Depending what decisions were made during development, additional driving modes such as 'eco' could be provided,

FIGURE 10.8   The drive-mode selector in a modern EV.

in line with the efficient motor management and pedal tune and feedback quoted previously. The clear advantage in terms of usability is the simplicity in the 'go forwards' and 'go backwards' command. No matter how simple it looks, there is always room for misreading the situation and having to rely on a visual display of the drive mode to compensate for the ambiguity. An example of that is the long-standing argument about layout: traditionally in sequential gearboxes, the backwards movement of the level changed a gear up and reverse was at the front end of the stroke, following after first and neutral gear. From an ecological design perspective however, with only two modes, front and backwards movement, the layout should be 'D' to the front and 'R to the rear, with 'N' in the middle. Then again, such arrangement would alienate drivers coming to EV after long experience of automatic or sequential ICE gearboxes, and so forth.

Lateral control (steering, in practice) is one of the few major areas of driving ergonomics where an EV does not necessarily introduce direct changes compared with an ICE vehicle; steering is naturally supported electrically (electric power steering (EPS)), but the same applies to most modern ICE vehicles. Steer-by-wire (electronic steering control, without any mechanical links to the wheels) appears to be closer to EVs, however in practice it is as much of a challenge as it is in ICE vehicles. Torque-vectoring technology (see Burgess, 2011) appears to be easier to apply in EVs and promises improved steering response and better handling of the vehicle; once more however the end result is dependent on the HMI challenge of tuning the system successfully across driver input.

10.4.2.3   Secondary Controls (HVAC, Wipers, Charging Control)

Outside the basic vehicle control and controls, the EV is not far from any other modern vehicle. The subtle differences can be narrowed to the operation of the HVAC, the control of charging times and the feedback from wipers in use. The latter does not have any functional or safety consequence; however, it can be a source of annoyance: conventional wipers are too noisy for an EV. They can provide additional auditory feedback, although judging by the area of their effect (windscreen, direct/foveal vision), the usefulness of such feedback is highly arguable. Control of heating, ventilation and air-conditioning (HVAC) however, plays an important role to the energy management—and efficiency—of the vehicle. In the case of automatic HVAC, responsibility falls on the HVAC engineers to specify a system that incorporates human response to heat as described in the relevant Chapter 7 of this book. Achieving that, in combination with an intuitive control interface (see Pheasant, 1996; Sanders and McCormick, 1993), can moderate the use of energy from the battery and therefore enhance driving range. The same applies to vehicles with manual HVAC, although in that case the impact of variation within driver population is much more difficult to account for. Manual HVAC however, is becoming less common in modern vehicles and nowadays applies only to the most affordable versions of models in the highly motorized countries (US, Japan, EU).

Last but not least, control of charging times and duration; unlike the chemical energy of fossil fuels in an ICE, which has to be filled during a stop at the fuel station, an EV can be charged virtually anywhere there is some form of electrical infrastructure, anytime and for as long as the battery capacity allows. On the other hand, the flow of energy, even when plugged to a quick-charging station, is many times slower than when filling up at the fuel station. For that reason, a charging management facility is highly recommended. The driver should be able to determine when and for how long a vehicle is being charged while plugged to the grid. In addition, long-charging times induce the risk for theft or other malicious acts on the charging equipment (plug and cable) in the absence of the owner. It is therefore advised that some security countermeasure is in place against such eventuality.

From the side of the driver, the tasks noted above place additional demands on the tactical and strategic level of driving: drivers have to be responsible and plan for their motoring energy needs. The need for recharging strategy is inversely proportional to the availability of EV infrastructure (quick-charging availability); the less infrastructure available, the more demanding planning becomes. Until that infrastructure matches the infrastructure for ICE vehicles, planning demand will be a key difference between the two types of powertrain.

10.5   EV: MUCH MORE THAN A CAR

We have so far witnessed the major implications of EV motoring to the driving task; the EV however, has the potential to be much more than 'a car'. In the immediately previous section, HVAC control and charging characteristics of EV were discussed; the weaknesses witnessed in charging time and HVAC energy demands can be mitigated if a facility is in place (could be through the IVIS or through a telematics system) to control charging and HVAC—for example, the EV could be pre-heated when connected to the grid, so that HVAC impact on energy consumption en route is minimized. Similarly, charging could be programmed to take place during specific time windows (e.g., if special electricity rates apply).

The previous section discussed the impact of electric motoring on longitudinal control and the importance of pedal tuning to match driver input and provide the expected controllability. Within that framework, the influence of regenerative braking and energy recovery was discussed. Concurrently, another radical change appears to be materialising slowly; acceleration control (both positive and negative) moves gradually to a single control—a pedal where depression prompts positive and release prompts negative acceleration. Already some 'gentle' braking drivers have such a braking behaviour (Gkikas, 2011) that they may virtually never have to use mechanical brakes in an EV. With the development of regenerative braking system capability, more drivers will reduce their use of the conventional brake pedal. This is potentially very important to the future of vehicle control ergonomics, as it essentially challenges the transfer of the ICE control layout—which is based on the legacy of the mechanical properties of the first automobiles—to the EV environment.

Looking at the bigger picture, EV is much more than a car, if the absence of EV-dedicated infrastructure is approached as a white sheet of paper for the development of energy-efficient technologies and utilities. An EV manufacturer has already provided an example use of the EV to power a house, by reversing the flow of energy through the charger back to the house. The typical amount of energy for a 60 mile/100 km trip is more than enough to power an average house in Europe for a day or two. Contemporary EV batteries have larger capacity than that. Even though this idea does not make sense as a permanent solution—the EV remains a vehicle and needs its own energy to carry its passengers around, it exhibits the potential of the EV as a mobile energy storage platform, which can be used in times of need.

REFERENCES

About, 2012. *History of Electric Vehicles-Early Years.* Accessed online: http://inventors.about.com/od/estartinventions/a/History-Of-Electric-Vehicles.htm.

Allen, T. M ., Lunenfeld, H. and Alexander, G. J., 1971. Driver information needs, *Highway Research Record, 366*, 102–115.

Appleby, A. J., 1988. *Fuel Cell Handbook.* New York, NY: Van Nostrand Reinhold.

Birrell, S. A., in press. Evaluation of current market green driving advisors: Smartphone apps or value added services. In L. Dorn (ed.) *Driver Behaviour and Training, Vol. V.* Aldershot, England: Ashgate.

Burgess, M., 2011. Torque-vectoring. *Vehicle Dynamics International.* Accessed online: http://www.vehicledynamicsinternational.com/downloads/VDI_Lotus_Vector.pdf.

European Commission, 2010. *A European Strategy on Clean and Energy Efficient Vehicles.* COM 186. Brussels: European Commission.

Gkikas, N., 2011. *Ergonomics of Intelligent Vehicle Braking Systems: The Operational and Functional Foundation for Driver-Centred Braking Systems.* Saarbrücken: Lambert Academic Publishing.

Gkikas, N., in press. Driving in the ear of IVIS and ADAS. In L. Dorn (ed.) *Driver Behaviour and Training, Vol. V.* Aldershot, England: Ashgate.

Keljik, J. J., 2009. *Electricity Four.* New York, NY: Delmar.

Lee, J. D., 2005. Driving safety. In R.S. Nickerson (ed.) *Reviews of Human Factors and Ergonomic, Vol. II,* pp. 173–218. Santa Monica, CA: Human Factors and Ergonomics Society.

Lee, J. D., and Strayer, D. L., 2004. Preface to a special section on driver distraction. *Human Factors, 46*(4), 583–586.

Michon, J. A. 1985. A critical view of driver behaviour models; what do we know, what should we do. M. L. Evans & R. C. Schwing (Eds.), *Human Behaviour and Traffic Safety,* 485–520. New York: Plenum Press.

Michon, J. A. 1993. *Generic Intelligent Driver Support.* London: Taylor & Francis.

Motlagh, A. M., Abuhaiba, M., Elahinia, M. H. and Olson, W. W., 2008. Hydraulic hybrid vehicles. In M. Kutz (ed.) *Environmentally Conscious Transportation,* pp. 191–210. Hoboken, NJ: John Wiley and Sons.

Pheasant, S., 1996. *Bodyspace.* London: Taylor & Francis.

Sanders, M. S. and McCormick, E. J., 1993. *Human Factors in Engineering and Design, 7th ed.* New York: McGraw-Hill.

Summala, H. 1996. Accident risk and driver behaviour. *Safety Science,* 22(1–3), 103–117.

Young, M. S., Birrell, S. A. and Stanton, N. A., 2011. Safe driving in a green world: A review of driver performance benchmarks and tehnologies to support "smart" driving. *Applied Ergonomics, 42* (4), 533–539.

Author Index

Subject Index